MATHEMATICS *and Science*

MATHEMATICS

and Science

edited by

Ronald E. Mickens
Department of Physics
Clark Atlanta University
USA

World Scientific
Singapore • New Jersey • London • Hong Kong

Published by

World Scientific Publishing Co. Pte. Ltd.
P O Box 128, Farrer Road, Singapore 9128
USA office: 687 Hartwell Street, Teaneck, NJ 07666
UK office: 73 Lynton Mead, Totteridge, London N20 8DH

MATHEMATICS AND SCIENCE

ISBN 981-02-0233-4

Printed in Singapore by JBW Printers & Binders Pte. Ltd.

Preface

The spectacular successes of the sciences is due in large measure to the use of mathematics in the creation and analysis of models for the phenomena of interest. Clearly, mathematics is a very effective tool/language for the formulation of scientific theories. Why is this the case? The question has been considered by philosophers, mathematicians and scientists since the dawn of human understanding and appreciation of mathematics, and the study of natural systems.[1] Mathematics seems to be an intellectual activity separate from the sciences. Its postulates may come from the processes of pure thought or may be suggested by experience. In either case, once formulated, they form a basis for which the structure of a mathematical system can be developed independently of their genesis.

In recent times, the essay of the physicist Eugene P. Wigner sparked renewed interest in this problem.[2] He asserted that "the unreasonable effectiveness of mathematics" is a mystery whose understanding and solution is yet to come:

> The miracle of the appropriateness of the language of mathematics for the formulation of the laws of physics is a wonderful gift which we neither understand nor deserve. We should be grateful for it and hope that it will remain valid in future research and that it will extend, for better or for worse, to our pleasure even though perhaps also to our bafflement, to wide branches of learning. (p. 14)

The purpose of the present book is to present the opinions of a number of distinguished individuals who have given careful thought as to why mathematics is so "unreasonably effective" when applied to the analysis of the natural sciences. (A very broad view is taken as to what constitutes the "natural" sciences. Included are the physical, biological and social sciences.) Each essay discusses one or more of the following questions:

1. What is the nature of mathematical truth?
2. Why is mathematics the language of science?
3. Is mathematics a science?
4. What properties does an area of knowledge have to possess before it is considered as being a science?
5. How has mathematics influenced the direction, interpretation and analysis of (particular areas of) science?
6. How has science influenced the direction of mathematical research?

It is not to be expected that the essays of this volume will settle, once and for all times, the "problem" of the effectiveness (unreasonable or not) of mathematics in the sciences. However, each essay, in its own right, provides a valuable contribution to a better understanding of the various issues of relevance for this topic.

The essays of this volume represent the personal reflections, thoughts and styles of the various contributors. Consequently, as editor, I have made only trivial changes to the manuscripts as I received and reviewed them. I hope that readers will find its contents as interesting and stimulating as I do. At the end of this volume is a Bibliography. It contains a short list of books and papers that give further discussions on the relationship between mathematics and the sciences. I would like to thank Dr. K. K. Phua, Editor-in-Chief of World Scientific Publishing for his enthusiastically endorsing this publishing project. I also wish to thank my wife, Maria Mickens, for not only her typing responsibilities related to this book, but, also, for her understanding and patience.

RONALD E. MICKENS
May 1990
Atlanta, GA

References

1. J. V. Graniner, *Math. Magazine* **61** (1988) 220–230.
2. E. P. Wigner, *Commun. Pure Appl. Math.* **13** (1960) 1–14.

Contents

MATHEMATICS *and Science*

On the Effectiveness and Limits of Mathematics in Physics

A. O. Barut
Department of Physics
University of Colorado
Boulder CO 80309, USA

1. On the "First Principles" of Physics and Mathematics

> *"It does not seem likely that the first principles of things will ever be adequately known ... "*
>
> *Voltaire*

In trying to understand why the laws of physics are most effectively formulated in mathematical form and what the limitations of this formulation are, we must look at the origins and the first principles of these two disciplines.

One may have the impression that physics and mathematics follow from well-established, well-defined rigid first principles. This is certainly not true for physics, and perhaps surprisingly, also not true for mathematics. Both the effectiveness and the limitations of physics and mathematics as applied to natural phenomena can be grasped from the realization that the basic entities underlying physical (and mathematical) theories are relative notions and strictly speaking undefined. Physics does not start from the beginning, but somewhere in the middle. The middle is the sphere of human experience and human experimenting with natural phenomena. There is necessarily an anthropic element at the beginning of physics. The same is true for mathematics. From the counting with one's fingers to numbers and from the shapes we observe to the geometric forms. From these anthropic beginnings we extrapolate in all possible directions.

When we do start somewhere in the middle, we begin by giving names to things and try to classify them, to find relations between things and to order them in space and time relative to us, the observers. And when we name things, it does not mean that we know these objects completely.

Physics is about partial aspects and partial relations of things which we only understand partially. The amazing thing is that in spite of all our ignorance, we can make rather precise statements and predictions about certain aspects of phenomena.

Imagine a huge complex machine, so complex that it would be impossible for an observer to describe all its external and internal parts, let alone to know its purpose and function. But he can easily discern a certain piston performing a periodic motion, for example, he can write a differential equation for its motion, measure its constants and make predictions about its further behavior. When I have to explain the existence of natural laws, or think about what is solid, precise and well-established in physics, I go back and begin with periodic phenomena, the sunrise every morning, and hence the definition of time. This is a beginning in the middle. We do not understand the sun, the light, the eye. But all these things do not matter at all for the observation of sunrise every morning.

We have the ability to abstract and isolate certain relevant features of phenomena and forget about the rest of the universe, or replace all the effects of the environment by simple effective agents. Thus as far as the motion of planets is concerned, they can be replaced in a first approximation by mass points moving in space. I call these first abstractions as simple observations, or simple facts. Next we have the more mysterious ability of reasoning, trying to connect and correlate different simple observations with each other in front of the Big Machine, assuming that there are connections.

The connections between different simple observations may be in the form of repetitions. We then count them and have invented the integer number system. Often they are repetitions with variations of the theme. Other times there are similarities, analogies, extensions. Yet we recognize regularities in change. Without going too much into the details and history, it is clear that this leads to the notions of independent variables to be varied, and other dependent variables, or functions. Thus the fundamental notions of mathematics, numbers, spaces and functions (or mappings) can be said to have their origin in physics.

Now the abstractions and extensions continue with different emphasis for physics and mathematics. Physical theories arose when it was realized that a certain subclass of phenomena could be combined into a single simple system, such as the solar system of Copernicus. An early example of a mathematical theory, the Euclidean geometry, is an abstraction, but also a valid physical theory, a system of axioms for the congruences of shapes in three-dimensional space under rotations and translations, in agreement with everyday experience.

Physical and mathematical theories transcend the immediate simple observations. They contain new elements or new notions in order that a logically closed world system can be constructed. They also imply new consequences to be verified if the system corresponds to truth. For physics we obtain a partial picture, a model for a part of the universe. In mathematics, this leads to a system of axioms about mathematical entities; numbers, spaces of objects and functions. The geometry of the triangle gave us the irrational number $\sqrt{2}$. The solution of algebraic equations leads to complex numbers. The equation $x^3 - 12x - 14 = 0$, for example, has three real roots. But according to Cardano's "casus irreducibilis" (1550), these real solutions are obtained by a path into the complex numbers. If we set $x = (u+v)$, our equation can be solved by $u^3 + v^3 = 14, uv = 4$ so that u, v are solutions of the quadratic resolvant $\xi^2 - 14\xi + 64 = 0$, or $\xi_{1,2} = 7 \pm \sqrt{-15}$. A more physical example, may illustrate the occurrence of complex numbers. The damped oscillator

$$\ddot{x} + 2\lambda\dot{x} + \mu x = 0$$

has both a periodic solution with an exponentially decreasing amplitude $x = Ce^{-\lambda t}\sin\omega t$, $\omega = \sqrt{\mu - \lambda^2}$, and a completely aperiodic solution $x = C'e^{-\lambda t}\sinh\omega' t$, $\omega' = \sqrt{\lambda^2 - \mu}$. Both of these solutions can be combined into a single solution if we allow the frequency ω to be complex.

We come now to the way we enunciate physical laws. We must first use the common language to describe phenomena. We can talk about our observations, and we can write about them. It is probably a common experience that writing, a conscious effort to put down and explain logical connections, takes us much further than talking. This includes inventing new words and new modes of language. This is how poetic or literary "laws" are formulated, if there are any, or rather there are so many of them.

The basic observations of physics have been quantified by measurement. This is done by calibration, by a choice of scales or units. The existence of periodic phenomena gave us a unit for time. We can use time and motion to define the unit of length (in the latest version, the distance travelled by light in a certain time interval), and finally to distinguish the motion of different bodies we introduce the unit of mass, the latter without completely knowing the true nature of what mass is, or is made of. It is remarkable that there are no other units in the rest of physics. So we have added numbers to our common language. But this alone does not go far enough. Although Newton's laws of dynamics can be expressed in words, it would be difficult to deduce all of its consequences in words. Newton in his Principia used the geometry as his language. The infinitesimal method introduced right then is much more powerful. It incorporates cause and effect and evolution in a quantitative way.

The mathematical language is, like any other language, for the purpose of communication. The common language is too vague, has redundancy. We require a precise language to mean exactly the same thing to everybody and to the experimenters who can repeat and compare their results to greater and greater precision. The literary language, on the other hand, may better appeal to the subjective experiences of the poet. Starting from some not very well-defined words (i.e., starting from the middle), other "derived words" are however exactly defined, for example velocity as the derivative of the position function with respect to time. These operations are then given physical meaning which may not always precisely correspond to physical situations. So we create a mathematical system with well-defined rules and operations, some of which we identify with our physical models. For example, sets of elements with rules of combinations (products), and other operations on them (e.g. differentiation, integration). We have now entirely new elements of a language to communicate complicated patterns and evolutions in a compact form: functions like $\sin \omega t$, $e^{-\lambda t}$, or $\sinh x$, in space and time; differential equations for cause and effect; a collection of sets, and so on. This new language is called mathematical knowledge, or mathematical truth.

Although the origins of mathematics are in physics, it has always been serendipitous. One structure leads to another — as the example of complex numbers mentioned above shows — by analogy, by generalization. So it is understandable how mathematics can have its own directions of developments into new domains from its own internal consistency dynamics, not necessarily connected to physics, even though some of these new domains find again applications in physics somewhere, some time. It is a different abstraction of the human mind, reasoning about the structure of the mathematical language itself.

Physical abstraction is more restrictive. We have to construct world systems congruent to, or approximating the natural phenomena both for their material existence and also for their functioning in the present, future, past, and in the small and in the large.

These common features of physics and mathematics, both products of human reasoning to go beyond the simple observations, seem to me the reason why the language of physics is mathematical and why it is an effective language. They are basically two faces of the same effort.

2. Progress and Limitations

I shall now return to the limitations of physics and mathematics in the quest for a complete description of nature.

The limitations of physical theories are two-fold. In the first place, the system or the theory, built to combine into one a whole set of simple observations (or primitive events), turns out to be always approximative. The system gives the main features but leaves a "fine-structure" out. Secondly, the primitive events themselves use approximative, undefined or unknown entities, whose internal nature has to be refined as we go along. For example, the primitive entity "electron" has undergone at least half a dozen evolutionary descriptions, and we still do not completely know what an electron is.

It may be argued that there is a certain physical knowledge about nature which is not approximative and seems to be absolute and exact. This concerns the symmetry principles and conservation laws. It is another place where physics and mathematics meet, deal with the same subject and arrive at the same results. It can be said that the true physical knowledge consists in recognizing the permanencies in phenomena in the midst of complex or chaotic appearances. Symmetry expresses the fact that the phenomena will appear the same to different observers at different times at different places. If this was not so, physicists would always have to publish, together with their experimental results, the position, orientation, the speed of their laboratory with respect to fix stars, the time, and so on. It would be a much more complicated world. The mathematical language to express the symmetry is group theory or geometry. We have learned since Felix Klein that to every group of transformations there corresponds a geometry, as the Euclidean geometry corresponds to the group of translations and rotations. Crystal symmetry corresponds to the finite subgroups of the rotation group. For the propagation of light and fast particles we find it more convenient to use another geometry, the Minkowski geometry. The symmetry in the larger sense is not just visual spatial symmetry, but also involves the symmetry of dynamics, the symmetry of physical laws, or the symmetry in the whole space *and* time. In space and time the world does not become, but it is. These symmetry laws give us also the physical quantities which do not change, the energy and momentum of system, and their angular momentum, its electric charge, when they are separated from the rest of the universe, and so on. These laws constitute essential knowledge in a very concise, economic and powerful way as far as they go. Their limitations come again from the limitations of the underlying basic notions. Since we do not know the structure of space-time in the very small and in the very large, we can only use the conservation of energy-momentum in limited regions of space, for example. Furthermore, we can never completely isolate one system from the rest of the universe. There are other imperfections. Crystal symmetry in nature is never perfect; we do not know how to predict which substances will form which symmetry, which of the 230 point groups do actually occur in nature. Macroscopic symmetries contain macroscopic parameters which cannot be derived from an underlying microscopic dynamical theory.

3. Physics and Mathematics in the Theory of Elementary Particles

I would now like to analyze the present status of fundamental physics from the point of view of the effectiveness of mathematical formulation and its limitations.

The "atomistic" point of view having triumphed in this century the "ultimate" physical theory at present consists of a number of assumed "fundamental particles" and their interactions. These interactions would yield bound states, these in turn yield more complex new bound states, clusters (nuclei, atoms, crystals), eventually the bulk matter, exactly as Newton had visualized (this quotation has often been used, but it is so good and relevant that I give it again here):

"Now the smallest particles of matter may cohere by the strongest attraction, and compose bigger particles whose virtue is still weaker, and so on for diverse successions, until the progression ends in the bigger particles on which the operations of chemistry, and the colors of natural bodies depend, and which by cohering, compose bodies of sensible magnitude" (Isaac Newton, Optics).

This approach provides both a model for matter and its dynamics, its functioning, in the form of interactions. The question is then what are the fundamental particles, how do they interact, and are we able really, starting from these basic entities, to calculate and understand the physical world. After the advent of wave mechanics, the old abstraction of a "mass point", a line trajectory in space-time, had to be abandoned in favor of a field $\phi(x,t)$ to describe a particle. Here is an instance where the initial concept of "particle" needed further precision. The model of two points with a force field between them is too crude for more elementary constituents of matter although it was adequate for macroscopic objects.

The field concept is well-known from continuum mechanics and wave phenomena. The mathematics is very effective here. It seems that if physical phenomena are simple and isolated enough to have striking distinctive geometric forms then the mathematical language is powerful enough not only to describe them precisely but also to predict them from dynamical differential equations. For example, the characteristic oscillations with discrete frequencies of membranes, shells or wave guides. These are obtained from the wave equations with appropriate boundary and initial conditions. To a first approximation it does not matter what material the membrane or the shell is made of. We only need its geometric form. But again the "fine-structure" of the phenomena does depend on it and we have to refine the model.

Schrödinger's model of the behavior of the electron in an H-atom is modelled after an oscillating ball. A wave equation with the natural condition that the amplitude decreases exponentially at infinity explains in a remarkable simple way the discrete frequencies of the atom, again to a first approximation.

The model has been refined by including the relativistic Minkowski geometry of space-time on which the fields are defined, and by including the internal spin degree of freedom of the electron. This is analogous to taking into account some of the material properties of the oscillating shell. The resultant Dirac wave equation is a more accurate description of the behavior of the electron than the Schrödinger equation.

The electromagnetic field concept originates with Faraday and Maxwell, and culminates in the prediction of electromagnetic waves and electromagnetic origin of light. One can imagine them to be the oscillations of a medium, called the aether, which has to be a relativistically extensible medium, not a rigid one, since no preferential directional propagation has been detected. We visualize wave phenomena more readily when they are related to the motion of a medium. But now it is more customary to think of them to be new independent quantities existing in space not attached to a medium, or to be radiated from charged particles.

The idea of two or more fields interacting (replacing two-or-more particles inter-acting) stems from the motion of mass points in external electromagnetic forces. When mass points were replaced by Schrödinger or Dirac fields, it was natural to study their behavior in external electromagnetic fields which then led finally to the equations of the coupled Maxwell and Dirac fields, namely the quantum electrodynamics.

Quantum electrodynamics (QED) is one of the finest achievements of theoretical and mathematical physics because of its simplicity, because of its wide applicability and large number of consequences, and because of its extreme accuracy, hence effectiveness, when compared with experiment. I think therefore I should write down these equations just to show their form. We have two fields. A vector field $A_\mu(x, t)$ for the electromagnetic field, and a 4-component spinor field $\psi(x,t)$ satisfying the equations

$$\left(i\gamma^{\mu}\frac{\partial\psi}{\partial x^{\mu}} - m\right)\psi(x) = e\gamma^{\mu}\psi(x)A_{\mu}(x)$$

$$\Box A_{\mu}(x) = e\overline{\psi}(x)\gamma_{\mu}\psi(x)\,.$$

Here $\Box$ is the wave operator, γ^{μ} the 4×4 Dirac matrices with summation convention over the index μ. It is not necessary for the following discussion to know everything about these equations. But I give this as a prime example how physicists can condense the basics of perhaps 99% of all physical and chemical phenomena into a two-line of equations. These equations can be derived from a variational principles (symmetry) and under some conditions it is a unique theory. For it is important that we try to characterize precisely the conditions under which we would have a compeling unique description.

When applied to the simplest atomic problem, the H-atom, quantum electrodynamics revealed further fine structure than is contained in the Dirac equation alone, in agreement with experiment. And when applied to the magnetic moment of the electron, its accuracy compared to experiment is "one part in 10^{10}". Enormous effort has been spent over the last forty years, and the effort is continuing, to test quantum electrodynamics to higher and higher order of accuracy, both theory and experiment. For when we extrapolate and construct a theory we have no assurances that we found the correct final picture of how nature works. At stake is nothing less than the "truth", the search for an objective, verifiable, precise rendering of nature. Any discrepancy may make our model a mental construct. There is still a two-standard deviation difference between theory and experiment in the best of all tests, in the magnetic moment of the electron.

But what are the limitations again?

The first limitation seems to be of mathematical nature. How to solve the above equations of QED, rather its operator form [i.e., when $\psi(x)$ and $A_{\mu}(x)$ are operator valued functions in the Hilbert space of states], and give to these equations a consistent meaning. It has not been possible, despite enormous efforts to show that this theory, mathematically, makes sense, is finite and consistent. The successes mentioned above come all from perturbative expansions after a somewhat artificial elimination of infinities. This makes a lot of physicists unhappy that this best of all physical theories is beset with seemingly unsurmountable difficulties. In such unsettled situations it may be best to go back to the sources. Recently, the much simpler classical version of Eqs. (1), written for ordinary functions $A_\mu(x)$ and $\psi(x)$ rather than operators, have been studied which give the same successful results as the perturbation theory of quantized fields, but with a much simpler physical and mathematical picture, and without infinities.

The correct interpretation of QED-equations is connected with the physical nature of the basic entities, the electron and its electromagnetic field. And this itself is connected with the basic questions of the physical interpretation of quantum theory: (1) Whether the fundamental description of the electron is deterministic or probabilistic, (2) how do we reconcile the particle and wave aspects of electrons and photons, (3) is there an objective entity called electron, or is it something whose properties are determined after observations? These are questions hotly debated still today after more than 60 years of quantum theory. For the purpose of this essay they show that we have come to the limit of validity of our initial concepts. What is an electron and what is light? Naming an entity does not mean that we know exactly what it is. The limitations thus may be rather of a physical nature than mathematical, or both, for they go together as we have seen.

So what are the first principles of physics? There are none, or, we are not sure. The deeper we probe the fundamental entities, the more vague, elusive and undefined they become. I suppose this is the same situation in mathematics, the continuum hypothesis, the axiom of choice, the undecidability of certain propositions, and so on. Even such simple concepts, like length, one of our fundamental units, is an idealization or abstraction, when compared with the atomistic nature of matter. We cannot realize at the atomic distances a length. In actual observation a line becomes a fractal with a dimension different from one, and then dissolves into the atomic structure.

I have talked about quantum electrodynamics as the fundamental theory of the present day physics. What about the rest of physics?

When a theory or method is successful, it has usually a wider range of applicability. At least we try to use it elsewhere with or without some generalizations. So it was natural to try the methods of QED for the physics of the nucleus, the radioactivity and the new particles produced in cosmic rays and in accelerators. Do they teach us something new as how the world is made of, our Big Machine?

Two distinct observations are relevant to our topic concerning the development of particle physics in this century.

First the constitution of the nuclei of atoms seemed to show that this structure is not of electromagnetic type. Thus a single unified picture based on a single basic interaction did not seem to be sufficient. The world was more complex. Other fundamental new particles have been introduced besides the electron and the electromagnetic field. The proton, the neutron, and most spectacularly, the neutrino, which has been postulated from theoretical internal consistency considerations alone and experimentally discovered 25 years later. This tells us that there are some very general basic laws of physics built over the last 400 years which are true, like energy conservation, as we have seen earlier. Many other new particles have been identified. We were confronted with the puzzle of finding some new simple basic set of particles out of which all these hundreds of particles might be constructed, in the same way as the many states of atoms are built from just electrons and nuclei.

At the root of the scientific method and research lies the belief, proved to be true so far in the history of science, that the world is basically very simple, there is an underlying order in the seemingly complex phenomena, and our theories become simpler and more encompassing the deeper the level we go to.

This belief seems now to be shattered in particle physics. The foundations of the present day theories are extremely complicated. There are too many "fundamental objects" for no good reasons. In an effort to go deeper to find the simplicity all sorts of hypothetical particles have been introduced. After all, if the neutrino was first hypothesized and later found, why not for many other particles and radiations. As time went on and having gotten used to some of these hypothetical particles and having granted their validity, further underlying constitution of these hypothetical particles themselves was hypothesized.

It is very important in scientific research to know at a given time, what we accept as fully established, justified, almost beyond doubt, amply confirmed, in order to build on it. I have emphasized the careful building of model and theories from simple observations, and thorough testing of them, and the slow process of reexamining the basic entities towards first principles, if they are any, which we cannot know beforehand.

Perhaps for the first time in history, we seem to have abandoned this thoughtful process in particle physics. Some complex utterly hypothetical entities are postulated as first principles from some extrapolation or generalization of not so fully tested principles. There is no hope to ever seeing these hypothetical particles. The extreme version of this approach is the new string theory of particles, where structures, i.e., strings or membranes, of dimensions 10^{-33} cm, are hypothesized as first principles from which ordinary particles of dimension 10^{-13} cm would somehow follow as oscillations of these strings. It is unlikely that we can postulate the first principles of things with any degree of certainty beforehand. These objects are so removed from our experience so far that I doubt whether such an endeavor is properly part of science, or scientific method. It is a dangerous process. We must be very careful to distinguish science, from intellectual ideology. If we start saying that "I *believe*" these are the first principles of things ("this is the theory of everything" — as the proponents of string theory for example, say) and everything will follow from them but it may take 50 or 100 years then what is the difference of this, from saying that "I believe" there are these various Greek gods each one being responsible for one thing and they can capriciously decide and rule everything. At any given time, science has a large body of well-established facts and a few small new hypotheses to be carefully tested. If, on the other hand, we have a very large body of hypotheses but very few tested ideas, we have an ideology. Science was supposed to free us from the tyranny of belief. The new trend may take us a full circle back to the domain of beliefs.

I prefer an open ended science, theories which are carefully established, mathematically consistent and well-established in their admittedly limited domain of applicability, but with an eye towards improving and understanding better and better the basic entities. This is how we push the frontiers. Therefore I think that quantum electrodynamics badly needs a completion before any other hypothetical new theory is introduced. Nowhere else have we left a theory unfinished, inconsistent and partly understood. True that other classical theories, mechanics, electromagnetism, thermodynamics, have limited validity, but they are mathematically consistent and closed.

Quantum electrodynamics is so close to nature with its remarkable accuracy, yet so far away with its mathematical inconsistency and incompleteness that physics cannot afford to let it stand in this way. There is much more at stake. Over the years many indications and arguments have been collected showing that if we did fully understand QED in a nonperturbative way, we may also understand the rest of the particle physics, resurrecting the electromagnetic model of matter. This would free us from the tyranny of the whole host of hypothetical particles.

These considerations may show, I hope, how physics and mathematics have a common origin and a parallel development in spite of their being seemingly different. They must go hand in hand. As a physical theory is sterile without the language of mathematics, so is a mathematical framework, however clever, which is far away from nature.

Bibliography

For related topics see, A. O. Barut, *Physics and Geometry- Non Newtonian Forms of Dynamics* (Biblipolis, Napoli, 1989).

Why is the Universe Knowable?

P. C. W. Davies
Department of Physics and Mathematical Physics
The University of Adelaide
Adelaide, South Australia 5001

When Sir James Jeans proclaimed that 'God is a mathematician', he articulated the physicist's conviction that the fundamental laws of the physical universe are expressible as simple mathematical statements. Why this should be so has fascinated generations of scientists. I believe, however, that the efficacy of mathematics in describing the physical world is only part of a larger mystery, which is why the universe is knowable at all.

Most scientists take it for granted that the world runs according to certain definite, if yet only dimly discerned, principles. Part of the scientific programme is to infer those principles from experiment and observation, and thereby "explain" reality. There is an assumption that the main obstacle in achieving the goals of this programme is a lack of resources, but given enough time and money, there is said to be no problem in principle why any physical process need remain a mystery.

At first sight, this quiet confidence in the power of the scientific method appears to be misplaced. On casual inspection, the world seems dauntingly complex and utterly beyond comprehension. Few natural phenomena overtly display any very precise sort of regularity that hints at an underlying order. Where trends and rhythms are apparent, they are usually of an approximate, qualitative form. The idea that buried among the chaotic data of experience are hidden principles of an exact mathematical nature is far from obvious. Indeed, centuries of careful investigation of the world by Ancient Greek and Medieval thinkers failed to uncover any but the most trivial mathematical examples of an underlying mathematical order in nature.

The situation is well exemplified by the case of falling bodies. Galileo's observation that all bodies accelerate equally in the Earth's gravity is counterintuitive precisely because it is usually wrong. Everybody knows that a lump of coal falls faster than a feather. Galileo's genius was in spotting that the differences which occur in reality are an incidental complication caused by air resistance, and are irrelevant to the properties of gravity as such. He was thus able to abstract from the complexity of real-life situations the simplicity of an idealized law of gravity.

Galileo's work is often taken to mark the beginning of modern science, and in the three and a half centuries which have followed, the trick of abstracting idealized laws from the complexity of raw data has proved very successful, in fact, more successful than we have any right to expect. The essence of the success can be traced to the power of analysis. By this I mean the practice of isolating a physical system from the outside universe and focussing attention on the phenomenon of interest. In the case of falling bodies, such isolation might involve experimenting in a vacuum, for example.

The fact that this procedure works so well is something of a mystery. The world, after all, is an interconnected whole. Why is it possible to know something, such as the law of falling bodies, without knowing everything? More to the point, why is it possible to know *so much* without knowing everything?

If the universe were an "all or nothing" affair, there would be no science and no understanding. We could never apprehend all the principles of nature in a single grasp. And yet in spite of the widespread belief these days among physicists that all the principles will indeed turn out to form a coherent unity, we are nevertheless able to progress one step at a time, filling in small pieces of the jig-saw without needing to know in advance the final picture that appears upon it.

Part of the reason for the success of this step-by-step approach is that many physical systems are approximately linear in nature. In physics, linear systems are those in which the whole is merely the sum of its parts, and in which the sum of a collection of causes produces a corresponding sum of effects. By analysis, one can chop up complicated systems into simpler components. An understanding of the behavior of the components then implies, *ipso facto*, an understanding of the whole.

Much of the linear mathematics developed in the eighteenth and nineteenth centuries was directed towards the description of linear physical systems. Techniques such as Fourier analysis work so well for heat conduction, small vibrations, electromagnetic field processes and other familiar phenomena because these phenomena are all linear to a high degree of approximation. On the other hand, they all turn out to be nonlinear at some level. When nonlinearity becomes important, it is no longer possible to proceed by analysis, because the whole is now greater than the sum of its parts. Nonlinear systems can display a rich and complex repertoire of behavior and do unexpected things.

Generally speaking, a nonlinear system must be understood in its totality, which in practice means taking into account a variety of constraints, boundary conditions and initial conditions. Although these supplementary aspects of the problem must be included in the study of linear systems too, they enter there in a rather trivial and incidental way. In the case of nonlinear systems, they are absolutely fundamental in defining the character of the system.

A good example is provided by the case of the driven pendulum.[1] Imagine a ball suspended from a string and driven by a motor at the pivot. (This is a so-called "spherical" pendulum.) The driving force is periodic, and aligned along a fixed horizontal direction. The driving frequency is close to the pendulum's natural frequency. The pendulum will start to oscillate. To the extent that the amplitude of the pendulum's oscillation is small, the system is approximately linear, and the oscillations will remain in the plane of the driving force. Because of friction, transient motions will soon decay, and the pendulum will settle down to slavishly follow the driving force, oscillating at the driving frequency irrespective of what that is. Its motion is both regular and predictable: literally "as regular as clockwork".

For larger amplitude swings, nonlinearity becomes important. The behavior of the pendulum is then dramatically different. The nonlinearity couples the driving force to the transverse direction too, so the pendulum is free to swing out of the driving plane. Its behavior now depends critically on the exact value of the driving frequency relative to the natural frequency. For some values the pendulum starts to gyrate, following which the ball settles down to trace out an elliptical path at the driving frequency. For slightly different driving frequencies, the pattern of motion is very different indeed. The ball swings this way and then that, sometimes in a plane, sometimes tracing a clockwise path, sometimes anticlockwise. No regularity is apparent. Indeed, it may be demonstrated theoretically that the motion is truly random: the system is an example of what is known as *deterministic chaos.*[2]

The spherical pendulum example graphically illustrates how a very simple nonlinear deterministic system with only a very few degrees of freedom can nevertheless display dynamical behavior of extraordinary richness and complexity. It can do wholly unexpected, indeed unpredictable things. The transition between regular and nonregular motion depends very sensitively on the ratio of the driving frequency to the natural frequency. The latter is in turn determined by the length of the string, which is a fixed constraint in this system. In the linear case this constraint is merely an incidental feature, determining only how long the transients take to decay. But in the nonlinear case it is crucial, an integral part of the problem. That is what I mean by saying that in nonlinear systems, constraints and boundary conditions are fundamental in determining the system's behavior.

Many other examples of what might be called the holistic character of nonlinear systems have been discovered. These include nonlinear waves and solitons, self-organizing systems, and far-from-equilibrium chemical reactions in which coherent global behavior results, either in the form of synchronous pulsations or spatial patterns. In all these cases, an understanding of the *local* physics (e.g. the forces between neighboring molecules), while possibly necessary, is by no means sufficient to explain the phenomena.

The foregoing systems can be mathematized to a degree, but that all-important 'unreasonable effectiveness' begins to wane somewhat. Because of their holistic, boundary-dependent nature, each system tends to be unique. General theorems applicable to a wide class of similar systems are hard to come by. Moreover, a common feature of these systems is that they tend to possess singularities in their evolution equations. Physically, this leads to instability, loss of predictability and a breakdown of uniqueness in their dynamical solutions. The result can be what is known as a bifurcation: one solution splits into two branches, each representing physically very distinct states. At bifurcation points the fate of a system becomes indeterminate. It might, for example, leap spontaneously into a new state of greater (and totally unexpected) organizational complexity.[3]

Rather than discuss these additional topics, however, I should like to restrict myself to one aspect of nonlinearity, namely chaos, because it serves to illustrate not only the globally indivisible character of nonlinear systems, but also the complete breakdown of the effectiveness of mathematics as a detailed description of the physical world. In recent years there has been a growing appreciation that chaotic dynamical behavior is by no means rare. Indeed, it seems to be a general feature of nonlinear systems. Joseph Ford, an expert on chaos, expresses it thus[4]: "Unfortunately, non-chaotic systems are very nearly as scarce as hen's teeth, despite the fact that our physical understanding of nature is largely dependent on their study."

The point about a chaotic system is that its behavior is truly random. This statement can be made rigorous using a branch of mathematics known as algorithmic complexity theory.[5] Here one defines the complexity of a string of numbers by the length of the smallest computer algorithm that can generate it. Thus the binary string 0101010101010101010... has very low complexity, because it may be reproduced by the simple algorithm "print 01 n times". More precisely, the information content of the algorithm is much less than the information content of the digit string. Such strings are said to be "algorithmically compressible". By contrast, a random digit string is algorithmically incompressible: no generating algorithm exists with significantly less information content than the string itself. This definition captures the intuitive quality of a random string that it should contain no trace of a pattern. If any pattern existed, we could encode it in a simple computer algorithm. Thus, for a random string, there is essentially no easier way of generating the string than to display a copy of the string itself.

These ideas can be carried over into physics by regarding the digit strings as predictive sequences for the motion of a physical system. For example, in the case of the pendulum, one could assign 0 for clockwise motion and 1 for anticlockwise motion. When the pendulum goes chaotic, the resulting binary digit string will be random and hence algorithmically incompressible. Under these circumstances, mathematics is no longer of any use to us in providing a compact description of the physical world.

This simple example gets to the heart of what one means by "the unreasonable effectiveness of mathematics". To be effective, a description needs to be more compact than the reality it seeks to describe. If our mathematical description contains as much information as the system's behavior itself, nothing is gained: the theory degenerates into a sort of xerox machine. The requirement of algorithmic compressibility is implicit in the central role played by *prediction* in science. Mathematics is effective precisely because a relatively compact mathematical scheme can be used to predict over a relatively long period of time the future behavior of some physical system to a certain level of accuracy, and thereby generate more information about the system than is contained in the mathematical scheme to start with.

To take the example of a generally non-chaotic system, consider the motion of the planets in the solar system. These are described, with "unreasonable effectiveness", by Newton's laws. The input information content is that of the laws themselves (rather small) and the initial positions and velocities of the planets. The quantity of information contained in the initial conditions depends on the precision of the input data, i.e., how accurately the positions and velocities are specified (three decimal places? ten? a thousand?). The greater the accuracy the more the information content (the longer the digit string of data). But the point is that the output of the computation – the positions and velocities of the planets at later times – has the *same* level of accuracy (more or less) as the input data. To be more precise, the errors propagate only linearly through the calculation. Thus the amount of output information is vastly greater than the amount of input information. Therein, of course, lies the usefulness of the procedure. One might say that the output data has been algorithmically compressed – into Newton's laws plus the (relatively modest amount of) input data.

By contrast, in a chaotic system the errors propagate exponentially. They grow so fast that they soon engulf the calculation completely and all predictive power is lost. Improved prediction can only be obtained at the price of knowing the initial conditions to ever greater accuracy, i.e., by gleaning escalating amounts of input data. But in all cases the output information content is little greater than the input information content. So nothing is gained by attempting to predict the behavior of a chaotic system. The calculation can at best merely keep pace with the reality. The physical system is, in fact, its own fastest computer. You might as well give up and just watch the reality happen.

All science is really an exercise in algorithmic compression. What one means by a successful scientific theory is a procedure for compressing a lot of empirical information about the world into a relatively compact algorithm, having a substantially smaller information content. The familiar practice of employing Occam's razor to decide between competing theories is then seen as an example of choosing the most algorithmically compact encoding of the data.[6] Using this language, we may ask: Why is the universe algorithmically compressible? and Why are the relevant algorithms so simple for us to discover?

It is interesting to relate these questions about algorithmic compressibility to Wigner's original question about the unreasonable effectiveness of mathematics. To do so, one must make a very important distinction, which is often blurred in discussions about the efficacy of mathematics, between the mathematization of the *theory* and of the *solutions* to that theory. Writing down a theory in terms of simple mathematics does not itself constitute a simple *algorithm* for generating descriptions of physical processes. As we have seen, the dynamical equations for some nonlinear systems can be extremely simple, yet their solutions may be chaotic and hence not reproducible by any compact algorithm. The spherical pendulum is a case in point. The Hamiltonian for this system is extremely simple, yet its solutions are exceedingly complex.

This point is not restricted to discussions of chaos. The general theory of relativity (a theory of gravitation) is famous for the mathematical simplicity and elegance of its field equations. It provides compelling evidence for the 'unreasonable effectiveness' of mathematics. Yet those same equations are notoriously difficult to solve. Their nonlinearity severely restricts general methods of solution, and even after sixty years of investigation, only a rather small number of exact solutions is known. So from the point of view of actually using the theory, it is perhaps not all that effective.

So we really have two mysteries. The first is to understand why the fundamental laws of physics are expressible in terms of relatively simple mathematical objects. The second is to understand why, under so many circumstances of interest, the solutions of the equations encapsulating those laws are also mathematically simple. One could imagine a world in which the basic laws were a set of simple equations, but for which no simple solutions existed. The actual behavior of physical systems might then be impenetrably complicated. In that case it is doubtful if we would ever discover the power of mathematics to describe the underlying theory, for nature would possess no conspicuous regularities to attract our attention to this line of reasoning.

The fact that the world is *not* like this is, as I have already indicated, due in part to relative smallness of many nonlinearities. Thus in the theory of gravitation, the motion of the planets in the solar system can be well described by Newton's theory, which is a linear approximation to the general theory of relativity. In this case there *is* a simple algorithm for generating solutions to the theory, and the behavior of the solar system is (to within the linear approximation) very simple and elegant, the planets tracing elliptical paths, etc.

Similarly, many physical systems are chaotic, but on a relatively long time scale. For shorter durations the motion is reasonably simple and predictable, i.e., describable by compact algorithms. Thus the behavior of the spherical pendulum during, say, one tenth of a period, is tractably regular. Another example concerns weather forecasting. Over a duration of weeks the weather is probably chaotic, but a next-day forecast is quite feasible.

One could go on to ask *why* so many nonlinearities are small. Here the discussion gets very complicated and will depend on the detailed physics. In some cases the answer is that the various constants of physics, such as Newton's gravitational constant and the electromagnetic fine structure constant have the values that they do. At present, there is no explanation for where these constants of nature come from. They occur as undetermined parameters in the theories of the various forces, and must be fixed empirically. There is no obvious reason why these constants have to have the values that they do, and not some other values.

The fact that the nonlinearities are small explains why we are able to abstract from the mess of raw data the underlying laws of physics. The example has already been given of how Galileo was able to infer the law of falling bodies in spite of the interfering complication of air resistance. The world seems to be such that, at least in part, we are able to extract the signal of the underlying simple mathematical law from the noise of real-life experiments. That this is so already points to important and unexpected properties of the laws. Barrow has conjectured[7] that this extractability of the signal from the noise might reflect something analogous to "optimal coding" in Shannon's theory of information.

Of course, none of the foregoing explains *why* the world is mathematically simple, only *how*. To answer the why question one must step outside of science. Some scientists resort to anthropic explanations at this stage.[8] For example, it could be argued that for conscious organisms to evolve and remain extant for reasonable periods of time (e.g. long enough to develop mathematics) some level of stability is required in the world. Conscious organisms could not arise in a totally chaotic universe. Moreover, the very definition of consciousness implies mental processes, which in turn imply information processing and algorithmic compression. Indeed, the brain is a marvellous example of the application of algorithmic compression: vast quantities of sense data are incorporated into a simple mental model of the world with impressive predictive capability. If the world were not algorithmically compressible, the concept of consciousness would almost certainly be meaningless. In short, we could not exist in a world which does not possess at least some algorithmic compressibility. Hence, given that we exist, it is no surprise to find ourselves living in a world where mathematics works well, for we could not be observers of any other sort of world.

It is important to realize that the anthropic argument does not constitute an explanation as such. It might still have been the case that the universe was chaotic and without observers to note the fact. For anthropic reasoning to constitute an explanation it must be augmented with an assumption about an ensemble of universes. Imagine that, instead of a single universe, there exists a vast (possibly infinite) collection of parallel realities, each with slightly different laws of physics. The vast majority of these universes would not be algorithmically very compressible, i.e., describable by simple mathematics. But these universes would go unobserved because they would not provide the necessary conditions for observers to arise. Only in a small fraction of universes, namely those which are highly algorithmically compressible, would life flourish and mathematicians evolve to ask why mathematics was so unreasonably effective in their world.

I have to say that I do not favor such an anthropic explanation, partly because one could resort to similar arguments to explain just about anything. (Why bother with science at all when an anthropic explanation for almost any feature of the world can be cooked up?) Also, it seems to me that invoking an infinity of unseen universes merely for the purpose of explaining this one is the antithesis of the very Occam's razor the success of which is part of what we are trying to explain.

It appears, therefore, that we are left with a deep mystery. There is no obvious reason why the laws of physics *have* to be such as to generate a universe which is highly algorithmically compressible, or cognizable. The fact that they are of this form must surely be a property of fundamental significance. However, as we have no idea where the laws of physics have "come from", or how large is the set of alternative possible self-consistent laws, there is no way of quantifying precisely *how* remarkable the algorithmic compessibility and cognizability of the universe is.[9]

In addressing the question of why so much of nature is mathematical, it is important to realize that each generation has its own idea of what constitutes mathematics. The invention of the ruler and compass in Ancient Greece led to the flowering of geometry. The Greeks saw geometry everywhere: in engineering, in music, in astronomy. They found it so successful that they built a theory of the universe around it, and combined it with the theory of numbers into an elaborate numerology. This was the first hint of the power of mathematics in describing the world.

The invention of accurate clocks in Medieval Europe led to a change of emphasis. Time now assumed a central role in the parametrization of physical processes, and the notion of the continuous flux of time led directly to the formulation of the theory of fluxions – calculus, in modern parlance. Thus was the science of mechanics born. These two great advances were accompanied by associated changes in world view. The Greeks regarded nature as a manifestation of numerological and geometrical harmony, while eighteenth and nineteenth century scientists saw the universe as a machine.

In recent years another technological invention has led to a third phase in this sequence. The development of the computer has given rise to new branches of mathematics, such as algorithmic complexity theory, and to a new world view which regards the physical world not as a set of geometrical harmonies, nor as a machine, but as a *computational process.* (The fact that these changes are technology-driven is an interesting issue, but outside the scope of this essay.)

The relationship between the physical world and the abstract world of computation is an important one. At the heart of the latter lies the concept of a Turing machine – a universal mechanical procedure for computing mathematical functions. Turing's imaginary machine consisted of a memory tape which could be inspected by a processor, following which an operation could be executed, such as changing the information on the tape, or moving it one space. This simple sequential device addresses the issue articulated by David Hilbert of whether a purely mechanical procedure could be found which would be able to prove or disprove any mathematical theorem. As is well known, Turing showed that this is impossible. There exist noncomputable functions; more precisely, there exist mathematical problems whose solutions cannot be decided on a Turing machine in a finite number of steps. Similar conclusions about noncomputable functions were reached by Alonzo Church. The reason is related to the existence of the real number continuum, i.e., uncountably infinite sets.

In spite of these inherent logical limitations of Turing machines, it was conjectured by Church and Turing that a Turing machine is capable of simulating all physical processes that can be completed in a finite number of steps. As pointed out by David Deutsch,[10] and discussed by John Barrow,[11] the Church-Turing hypothesis addresses at a very deep level the relationship between mathematics and the physical world. Deutsch and Barrow invite us to imagine two black boxes, one containing a Turing machine, the other a physical system. From a given input these black boxes furnish a certain output, with the Turing machine simulating the physical process taking place in the other box. If the Church-Turing hypothesis is correct, then we could not tell from the output which box had furnished the answer, i.e., which box contained the real physical system and which the Turing machine.

In today's computer age we take it for granted that machines can simulate basic arithmetic operations such as addition and multiplication. Yet there is no known reason why the laws of nature are such that physical processes can do this. Hamming seems to be making the same point when he writes[12] of how the integers and simple arithmetic operations on them are so well adapted to many real-world situations (such as counting sheep): "Is it not a miracle that the universe is so constructed that such a simple abstraction as a number is possible? To me this is one of the strongest examples of the unreasonable effectiveness of mathematics. Indeed, I find it both strange and unexplainable."

It is interesting that the Church-Turing hypothesis cannot be correct in a Newtonian universe, in which energy is infinitely divisible. However, quantum mechanics ensures that finite systems possess discrete energy levels, circumventing this problem. Thus, the universe can only be regarded as a computational process so long as something like quantum, as opposed to classical, mechanics is correct. A quantum version of the Church-Turing hypothesis has been discussed by Deutsch.[10]

The issue of computability has an important bearing on the nature of physical theories. I have stressed the fact that one should make a sharp distinction between a theory and the algorithm that implements it (usually, distinguishing the dynamical equations from their solutions). It may be that some physical theories will need to be structured differently, so that this distinction is blurred. In which case there will be little difference between our initial mathematization of the theory and our mathematical implementation of it.

Robert Geroch and James Hartle[13] have argued that with the quantum theory of gravitation, and especially quantum cosmology, we may already be presented with such a situation. They point out that the theory can involve certain summations over compact four-dimensional spaces with different topologies, and then note a theorem which says that there is no computer program that can compare the topologies of 4-manifolds, print out 'same manifold' or 'different manifold', and then halt. This suggests that the theory predicts for the result of an in-principle measurable quantity a noncomputable number.

A theory which predicts noncomputable numbers for measurable quantities cannot be tested by any mechanical procedure. Supposing a technician measures the quantity to some accuracy ε, we cannot know whether this measured value lies within ε of the predicted value because the defining quality of a noncomputable number is that there is no mechanical procedure for computing that number to within known precision. And yet a predicted value certainly exists! Indeed, it could be extracted to a given level of accuracy non-mechanically, by a human being performing skilful and insightful mathematical manipulations. With each increase in accuracy, different techniques would be called for requiring still more skill and ingenuity, and so on *ad infinitum*.

Until now, I have been using the word computability to mean something that can in principle be computed in a finite number of steps. However, in the real universe, the number of permitted computational steps cannot be made arbitrarily large. There are fundamental limitations connected with the existence of the finite age of the universe and the finiteness of the speed of light. Yet the mathematical encodement of our physical laws employs mathematical objects that are defined in terms of idealized operations which demand unlimited computational power. Is this consistent?

Rolf Landauer[14] has argued that "not only does physics determine what computers can do, but what computers can do, in turn, will define the ultimate nature of the laws of physics. After all, the laws of physics are algorithms for the processing of information, and will be futile, unless these algorithms are implementable in our universe, with its laws and resources". He points out that we cannot, for example, distinguish π from a near neighbor because even if we could commandeer the entire cosmos as a computer it would have finite computational power. If one takes the point of view that this inherent *computational* limitation implies an inherent *physical* limitation on the universe (turning the usual argument on its head) then the curious situation results that this limitation will be time-dependent. The reason is that at earlier epochs, the universe was in a sense smaller, and its computational power more limited. Taking this to the extreme, one arrives at a picture in which the laws of physics are not even defined at the moment that the universe comes into existence, and that they gradually "condense out" of nebulous origins as the size, and hence computational power, of the universe grows. This speculation is along the line of Wheeler's "law without law" suggestion.[15]

The subject of quantum cosmology has another, and as far as I can see unrelated, implication for the thesis of this essay. I have been arguing that approximate linearity is important because it enables us to quasi-isolate parts of the universe and study them in great generality. The essential requirement here is actually not linearity as such, but what physicists call *locality.* This means that all the relevant factors which bear on a physical system originate in the physics of the immediate environment of that system. Thus we do not have to worry about the state of a distant quasar when experimenting with falling bodies on Earth, for example. Linearity is usually a necessary, but not always a sufficient condition for locality. Quantum mechanics is a linear theory, but involves some subtle nonlocal effects that have lately been much discussed, usually in the context of the so-called Einstein-Podolsky-Rosen (EPR) experiment, and the associated Bell inequalities.[16]

This is not the place to discuss the EPR experiment in detail. Suffice it to say that quantum states can be constructed in which two particles, e.g. photons, fly apart to great separation (in principle they could travel to opposite ends of the universe) and yet still constitute a physically indivisible system in the following sense. In quantum mechanics there is inherent uncertainty in the outcome of measurements. Therefore the result of a measurement performed on one particle of the pair is not determined in advance, even in principle. Nevertheless, because of the aforementioned nonlocality, as soon as a measurement has been performed on one of the particles the outcome of measurements performed on the other particle, in a far distant place, is instantly determined. That is, there is a linkage between the particles, over and above anything that can be accounted for as due to simple interaction. (Unless, that is, one is willing to countenance faster-than-light signalling between the particles, with all the attendant causal paradoxes which that entails.)

The EPR experiment is designed to manifest in a striking way the nonlocality inherent in quantum mechanics. Yet this nonlocality is a general feature of quantum systems. The mathematical description of a quantum system is contained in something called its wave function, which can be used to compute the relative probabilities of different outcomes of measurements. For example, the wave function of an electron is generally spread throughout space, indicating that there is a nonzero probability of finding the electron anywhere. Thus, although the electron is a point object, its wave function is nonlocal in nature. If the electron is trapped in an atom, then its wave function is rather sharply concentrated in the vicinity of the nucleus, which tells us that the overwhelming probability is that the electron will be found close to the nucleus. In this case the wave function is quasi-localized.

The curious nonlocal effects associated with the EPR experiment do not manifest themselves in daily life because the wave functions of macroscopic objects are exceeedingly sharply peaked around definite points in space. Why this should be so is a deep problem. It is the problem of how the classical world of experience emerges from the quantum world that is the true underlying reality. If it were not so then the world would not be analyzable into discrete, well-defined objects at definite locations in space.

The universe would form a manifestly indivisible whole, and any mathematical description would be hopeless.

Now many physicists believe that this was actually the state of affairs during the very early stages of the universe. That is, the universe came into existence as a quantum phenomenon.[17] Attempts have been made to write down a "wave function for the whole universe" which is intended to describe this quantum cosmos. The challenge is then to understand how, during the course of its evolution, this wave function became very sharply peaked around the classical state of the universe that we see today. The solution to this problem cannot be separated from the philosophical difficulties concerning the nature of observation in quantum mechanics, and the interpretation to be given to the notion of a wave function of the entire universe. It remains a major outstanding problem of physics, and one which has a direct bearing on the success of mathematics in describing the world.

Even given that the universe *is* both algorithmically compressible and in large part (at least quasi-) local, there remains a deeper mystery concerning why we, i.e., modern *homo sapiens*, are so "unreasonably effective" at discovering the relevant algorithms. Why is it that, at the everyday level of observation, the world appears such a hopelessly complex jumble of phenomena, yet with a fairly modest investment of effort (say fifteen years of education) so much can be explained in terms of mathematical principles? Why does nature require a nontrivial and yet entirely manageable amount of mathematics for the successful description of such a large part of it?

Again, we really have no right to expect this. One could envisage a world that was so conspicuously ordered that its principles would be transparent to everyone. Alternatively, one can imagine a different world in which the principles of nature defy all straightforward attempts at unravelling. Why should so much of nature be just sufficiently subtle as to be comprehensible to human beings after fifteen years or so of mathematical education? Or to recast the question in a slightly different form: Given the almost limitless length of time apparently available in the future for progress in mathematics, why is it that only a few centuries of mathematical endeavour are sufficient for dramatic advances in understanding the physical world? After all, a few centuries of work is a nontrivial investment, and yet minute in relation to what could be achieved throughout all the future history of mankind.

It might be countered that the question is circular. There is always the risk of circularity in questions involving the human beings that ask them.

(The circle becomes vicious and Russellian with, for example, the following: Why am I asking this question?) In the present discussion, the question: Why does a nontrivial yet potentially manageable mathematical education enable such enormous progress to be made in understanding nature? seems to amount to no more than asking: Why do I happen to be living at that historical epoch at which the mathematics appropriate to rapid advances in understanding physics becomes available? Obviously *some* physicists will witness that epoch, and they will be the ones who will ask such questions. The situation resembles the football pools winner who asks: Why me? The answer is, *somebody* has to win.

The appeal to circularity aims to define the problem away by rooting the origin of the question in the emergence of the very circumstances that the question addresses. Yet this evasion is altogether too slick. The mystery is actually less anthropic than it might at first appear. But let us grant for now that the issue really does boil down to one of epochs in human history. We can imagine that scientists (whether human or our organic or machine descendents) will endure in the universe for millions or even billions of years. Suppose some remote descendent in the far future plots a graph of the rate of fundamental discoveries in physics (quantified in some appropriate way) against time. There would seem to be two possibilities. The first is that these discoveries go on accumulating more or less steadily throughout history. In this case our own epoch would be distinguished by the fact that it is located exceptionally near the initial sharp rise in the graph. Alternatively, it might be that the pace of discovery reaches a maximum (about now) and then declines. In this case one notes with interest that we are exceedingly privileged to be living at just that epoch where the peak in the graph occurs.

If the first scenario is correct, it implies that the present-day science is not, in fact, as successful as is commonly supposed. Our currently rather limited capability in mathematics would be seen as having limited our understanding of nature to a very crude and superficial level. In fact, today's mathematics would not be "unreasonably effective" at all. Rather, its apparent effectiveness could be attributed to the fact that we have chosen to investigate only those systems for which this primitive level of mathematical expertise happens to work. Indeed, the vast majority of natural phenomena remain a mystery at our epoch. Possibly they always will, if the necessary algorithms are beyond either human resources, or even beyond human mental capacities altogether.

I wish to argue that the foregoing is not the case, and that the second scenario is the correct one. There are good reasons to suppose that during the last three hundred years Mankind has witnessed a unique sociological phenomenon. The pace of discovery of the fundamental principles of physics has been enormous, and some physicists are already predicting that a "Theory of Everything" is within our grasp. Such a theory would embody a description of spacetime and of all the particles and forces of nature in a single scheme, possibly even a single, simple formula. At present hopes are pinned on the so-called superstring theory,[18] which holds out the promise that a consistent quantitative description of all physical processes will flow from a simple Lagrangian.

It is often objected that the physics community has reached similar states of euphoria about a unification of physics in the past, only to find that the treasured edifice comes crashing down with some new discovery. What this rebuttal neglects, however, is that the old edifice is rarely wrong as such, merely slightly flawed. After all, Newton's theory of gravitation remains good within a wide domain of validity. Even if a totally satisfactory Theory of Everything eludes us, it will remain true that we have already almost certainly achieved entirely satisfactory mathematical descriptions of the four known forces, and of the various fundamental particles of nature. In addition we have a relatively detailed knowledge of the evolution of the universe at least from the first one second until today. Certainly it is easy to imagine that gaps in our knowledge in this area will be filled by observations made over the next century.

My conclusion is that the uncovering of the fundamental laws of physics is a remarkably transitory phenomenon in the history of human progress. I do not expect our descendents in a thousand, let alone a million years, to be piling up discoveries in this field at the same pace as us. Minor refinements there will undoubtedly be, but ours is the millenium which will go down in history as the one where suddenly the basic laws of the universe fell into mathematical place, at least to very good approximation. Richard Feynman has written[19]: "Some of my colleagues said that this fundamental aspect of our science will go on; but I think there will certainly not be perpetual novelty, say for a thousand years. This thing cannot keep on going so that we are always going to discover more and more new laws."

Part of the reason for my belief is that our current theories are enormously extrapolatable. Consider quantum electrodynamics (QED) for example, the theory of the interaction of photons and electrons. Taking into account its classical limit in Maxwell's equations, it has been tested over length scales ranging from extragalactic (10^{24} cm) magnetic fields to 100 GeV photons (wavelength 10^{-16} cm), where the theory shades into the new unified electroweak theory. Quantitatively, QED has been tested to ten figure accuracy using subtle atomic correction effects. This level of success is sufficient to encompass almost all electrodynamic processes of interest in the universe, with the exception of some very high energy cosmic ray phenomena.

Many theorists working on the new brand of unification theories, such as superstrings, believe we already have satisfactory theories of the weak and strong, as well as electromagnetic, forces up to the so-called unification scale of 10^{14} GeV energy, and possibly as far as the Planck scale (10^{19} GeV). If correct, this would take care of everything that had ever happened in the universe since the first 10^{-32} sec or so. The point here is that nature defines natural energy scales at 100 GeV and at 10^{14} GeV but that between those two scales there is no reason to suppose that existing theories will break down. As direct experimentation at 10^{14} GeV would demand particle accelerators the size of the solar system, it is probable that the physics of this ultra-high energy regime will not be explored within the remotely forseeable future, and possibly never. The planned superconducting supercollider will reach TeV energies (1 TeV = 1000 GeV). Unless this machine uncovers unexpected new physics, it is likely that experimental particle physics will have reached its culmination.

I should point out that I am making a clear distinction between the elucidation of the *fundamental principles* of physics, and a full understanding of all the messy details of particular physical systems. Thus it might be that we shall not understand, say, the physics of biological organisms for a long time in the future. I am assuming that the latter do not introduce anything *fundamentally* new. This could certainly be challenged, if complexity as such is a physical variable, and one entertains the possibility of emergent phenomena in physics.[20]

In spite of the latter possibility, I am convinced that we are living at a time of unique progress in our understanding of the physical universe. If I am right then it reflects an extraordinary fact about the mathematical basis of this understanding. In the three short centuries during which this advance has taken place the physicist's mathematical requirements have ranged from the calculus to certain current topics in topology and group theory needed for the superstring theory. In fact, the superstring theory is ahead of the mathematics just now, but Witten, for example, forsees the development of the necessary mathematics as taking place over the next fifty years.[18] If so, then three and a half centuries of mathematical progress will be enough to provide what looks to be an entirely adequate description of fundamental physical processes that is likely to remain substantially unmodified for many millenia, if not for ever. If God is a mathematician, he seems to be a twenty-first century mathematician.

I find this very curious, because although fundamental physics may be approaching quasi-culmination, mathematics seems open-ended. One can envisage that mathematical theory will go on being elaborated and extended indefinitely. How strange that the results of just the first few centuries of mathematical endeavour are enough to achieve such enormously impressive results in physics.

References

1. D. J. Tritton, *Ordered and Chaotic Motion of a Forced Spherical Pendulum, European J. Phys.* **7** (1986) 162.
2. For a review of chaos, see J. Ford, *What is Chaos that We Should be Mindful of It?*, in *The New Physics*, ed. P. C. W. Davies (Cambridge University Press, 1989).
3. For a review of self-organizing systems, see I. Prigogine and I. Stengers, *Order Out of Chaos* (Heinemann, 1984).
4. J. Ford, *How Random is a Coin Toss?, Phys. Today*, April (1983) 4.
5. G. Chaitin, *Algorithmic Information Theory* (Cambridge University Press, 1988).
6. R. J. Solomonov, *A Formal Theory of Inductive Inference. Part I, Information and Control* **7** (1964) 1.
7. J. D. Barrow, *The World Within the World* (Clarendon Press, 1988) 292.
8. J. D. Barrow and F. J. Tipler, *The Cosmological Anthropic Principle* (Clarendon Press, 1986).
9. P. C. W. Davies, *What Are the Laws of Nature?*, in *The Reality Club* 2, ed. J. Brockman (Elsevier, 1990).
10. D. Deutsch, *Quantum Theory, the Church-Turing Principle, and the Universal Quantum Computer, Proc. Roy. Soc.* **A400** (1985) 97.
11. J. D. Barrow, *The World Within the World* (Clarendon Press, 1988) 264.
12. R. W. Hamming, *The Unreasonable Effectiveness of Mathematics", Am. Math. Monthly* **87** (1980) 81.
13. R. Geroch and J. B. Hartle, *Computability and Physical Theorems*, in *Between Quantum and Cosmos*, eds. W. H. Zurek, A. van der Merwe and W. A. Miller (Princeton University Press, 1988) 549.
14. R. Landauer, *Computation and Physics: Wheeler's Meaning Circuit?*, in *Between Quantum and Cosmos*, eds. W. H. Zurek, A. van der Merwe and W. A. Miller (Princeton University Press, 1988) 568.

15. J. A. Wheeler, *Beyond the Black Hole*, in *Some Strangeness in the Proportion*, ed. H. Woolf (Addison-Wesley, 1980) 339.

16. For a popular review, see P. C. W. Davies and J. R. Brown, *The Ghost in the Atom* (Cambridge University Press, 1986).

17. For a popular account, see S. W. Hawking, *A Brief History of Time* (Bantam, 1988).

18. For a nontechnical introduction, see P. C. W. Davies and J. R. Brown, *Superstrings: A Theory of Everything?* (Cambridge University Press, 1988).

19. R. P. Feynman, *The Character of Physical Law* (B. B. C. Publications, 1966) 172.

20. P. C. W. Davies, *The Cosmic Blueprint* (Heinemann, 1987: Simon & Schuster, 1988).

Mathematics in Sociology: Cinderella's Carriage or Pumpkin?

Patrick Doreian
Department of Sociology
University of Pittsburgh
Pittsburgh, PA 15260, USA

Reading the discussion of the "unreasonable effectiveness" (Wigner, 1960; Hamming, 1980) is an ironic tonic. Rather than viewing mathematics as a wonderful and useful gift, most sociologists regard it, at best, as irrelevant or see it as a spectre or curse. Some exhortations to use mathematics in sociology sound as if mathematics is a magnificent carriage that will, in the end, be as effective in the social sciences as it is in the physical sciences. We know the glittering carriage took Cinderella to a world beyond her dreams. We also know that pumpkins are singularly useless for travel and many sociologists would, in this imagery, see mathematics as a stationary vegetable. Any claims to revolutions would be viewed as thinking in circles.

This essay will issue no clarion call. Instead, it is a reflection on using mathematics in sociology and is structured as follows:

1. some remarks on sociology;
2. a look at scientific revolutions;
3. some considerations of the distinction between mathematical models and quantification;
4. some reasons for the relative absence of mathematics in sociology; and
5. an assessment of the chances for change.

1. Remarks on Sociology

There are, alas, many definitions of sociology. Some focus on the analysis of social action and its meaning for social actors, some on the analysis of social relationships and their consequences, and some on social structures and processes. Lenski and Lenski (1978:4) define sociology as "the branch of modern science that specializes in the study of human societies". This is adopted here even if many sociologists (often explicitly in their ideology, or implicitly in their behavior) deny the claim to science.

While the intellectual roots of sociology go back to the Enlightenment, and even to thinkers and philosophers of antiquity, the term itself was coined by Comte (1855). Trained as a mathematician, physicist and engineer, he proposed a classification of the sciences in an analytical and historical hierarchy of decreasing generality and increasing interdependence and complexity (Cohen, 1985). Comte saw mathematics as the most general of the sciences, forming the foundation of the hierarchy. The natural sciences came next before the final science of sociology. Clearly, in his view, there is nothing alien about mathematics in sociology. Nor is it clear that any of the founding thinkers of the field held to explicit exclusion of the use of mathematics.

Yet it is abundantly clear that the field overwhelmingly eschews the use of mathematics: some are simply non-mathematical while others are anti-mathematical. Most sociological books are devoid of mathematics. Indeed, a 700-page history of sociological analysis (Bottomore and Nisbet, 1978) makes no mention of it. In addition, sociology "is characterized by a proliferation of overlapping and sometimes competing "schools of thought" and, consequently, there is great disagreement among sociologists about the fundamental problems, concepts, theories, and methods of research in the discipline" (Smelser, 1967:2). It may well be that it is this disagreement that permits a small group of scholars to lay some claim to "mathematical sociology". Loosely, mathematical sociology is the study of human societies using the language(s) of mathematics. A stricter sense of the term is the use of mathematics to create generalizing and cumulative sociological theory.

In practicing the craft of mathematical sociology, most would agree with the general scheme outlined by Olinick (1978). (See also Roberts (1974).) The modeling cycle involves an abstraction from "the real world" to a mathematical system or model.

Within the mathematical system, logical argument (deduction) is used to derive mathematical conclusions which then can be interpreted in the light of information generated from the real world. The notion of an "experiment" is extended to include some form of observation of the non-experimental real world (in much of its complexity). The steps in the modeling cycle that are seen as most vulnerable are the abstraction to a mathematical model and the limitations on real world data that are needed to provide a viable interpretation given the mathematical derivation contained within the model.

Those championing the cause of using mathematics in sociology can point to a large set of examples of, it is claimed, successful use of mathematics (see below). There are dedicated journals (for example, *J. Math. Socio.*) and there are subfields heavily using mathematics, for example demography, game theory, and decision theory. Yet it is impossible to escape Sørensen's (1978:367) summary conclusion that mathematical sociology remains a subfield and "some may find it a fairly esoteric enterprise with little impact on sociology in general". Hayes (1984) concurs, finding it a struggling sub-specialty with minimal impact on social theory. Realizing Comte's vision will require nothing less than a scientific revolution.

2. Scientific Revolutions

Kuhn (1964) proposed a theory of scientific change as a dynamic of alternating periods of revolutionary change and "normal science". Normal science is practised within a shared paradigm that is composed of shared knowledge, shared theories, shared methods and models, and a common form of explanation. Such normal science is interrupted by revolutions that imply sharp breaks, at least in some respects, with the period prior to the revolution. Contrary to his initial formulations, not all revolutions arise from periods of crisis where the paradigm breaks down in the face of cumulated results. And his (profoundly sociological) analysis fares far better with the physical sciences than with the biological sciences. Revolutions may also vary in scope. Yet as Cohen (1985:27) points out, "Kuhn's analysis has the solid merit of reminding us that the occurrence of revolutions is a regular feature of scientific change and that a revolution in science has a major social component – the acceptance of a new paradigm by the scientific community". It is not clear the extent to which the succession occurs by persuading other scientists to change their views, or whether the mechanism is demographic in the sense that the previous paradigm adherents are replaced by the next generation. In either case, there is a transition.

Revolutions are easier to recognize with hindsight and much has been learned from the Copernican, Newtonian, and the Einsteinian revolutions, or the plate-tectonic revolution in geology. Cohen (1985:40–44) suggests ways of perceiving if a revolution has occurred:

1. the "testimony of witnesses", i.e., the judgment of the individual scientists and non-scientists of the time;
2. examination of the documentary history of the subject in which the revolution was said to have occurred;
3. the judgment of competent historians; and
4. the general opinion of working scientists in the field (after the alleged revolution).

The evidence, among other things, has to point to the accumulation of factual data and the nature of scientific theory as it changes dramatically.

The physical sciences have had their revolutions, meeting all of these criteria. Concerning the role of mathematics in such dramatic changes, Kline (1985: 216) writes that mathematics "has furnished the marvellously adept Euclidean geometry, the pattern of the extraordinarily accurate heliocentric theory of Copernicus and Kepler, the grand and embracing mechanics of Galileo, Newton, Lagrange, and Laplace, the physically inexplicable but widely applicable electromagnetic theory of Maxwell, the sophisticated theory of relativity of Einstein, and much of atomic structure. All of these highly successful developments rest on mathematical ideas and mathematical reasoning".

It is abundantly clear that on all four criteria – except perhaps the first, for a short period after World War II – that no revolution has occurred in sociology, let alone one carried by mathematics. No statement like Kline's is possible for sociology today, although there was a time when it seemed close. Fueled by the striking successes of game-theory, cybernetics, and operations research it seemed that mathematical social science would finally come of age. Yet Wilson (1984:222) writes "... though the goal of mathematical social science has been pursued for two centuries with considerable energy and talent, the results have been disappointing". Hayes (1984:325), noting the great optimism of thirty years ago reflected, in part, by the idea that a critical mass had finally formed, concludes "the relationship between mathematics and sociological theory seems today more ephemeral than ever". His conclusion, despite notable accomplishments, is that "mathematical sociology ... appears today as a rather disorganized array of models and strategies".

It can be argued that asking whether there could be such a revolution with respect to mathematics and sociology is entirely premature. If sociology is in a pre-paradigm state – many competing theories, approaches, methods with little evidence of anything shared – how could there be a scientific revolution in the sense of a paradigm shift? We still can argue that mathematics has failed to create such a sociological paradigm – along with all other contending subfields. However, there is one area where mathematics has been mobilized extensively, often implicitly. Statistical data analysis is widespread and rests on quantification of some sort that seldom leads to mathematical modeling.

3. The Distinction between Mathematical Models and Quantification

Lave and March (1975:52–73) capture fully the spirit of (mathematical) modeling: formulating models of social processes in a disciplined, creative and playful fashion. The social theory is directly translated into a mathematical model (compare Fig. 1): The abstracted process model is in mathematical form. Lave and March also discuss truth and beauty as criteria for evaluating mathematical models. Under beauty, there is a special emphasis on simplicity (parsimony – something sociologists tend to find very hard to accomplish), fertility (generating many interesting predictions per assumption, (c.f Heckathorne, 1984)), and surprise (generating something we did not expect, i.e., was not immediately obvious from the initial assumptions). Good mathematical models are abstract, rigorous, and fertile. They do exist, but at the margins of the field.

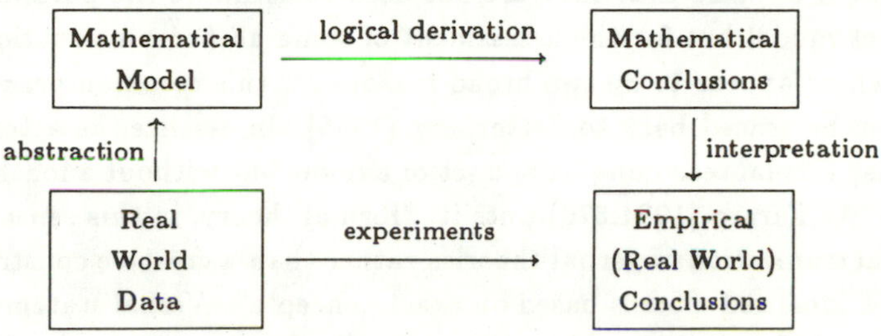

Fig. 1. Schematic diagram of a mathematical modeling cycle.

The argument against mathematical models is frequently mounted as an attack on quantification, a practice finding extensive use in the statistical analysis of data. There are, of course, many qualitative mathematical models (for example those using graph theory, algebraic topology, or modern algebra) so that even if the attack on quantification is successful, a whole class of viable (mathematical) models would remain. Clearly, statistical data analysis can be mobilized to estimate parameters of a mathematical model. As such, it is a technical procedure applied for a particular instantiation of a process being modeled. In terms of Fig. 1, the quantification is a process solely in the mapping contained in the real world part of the modeling process. It has little to do directly with the abstracting and interpreting homomorphic mappings, or the deduction within a mathematical model. It can be guided by them but, if done incorrectly, can render the abstraction, deduction, and interpretation as worthless. Quantification is measurement, while model building is the construction of social theory in mathematical form.

Data analysis rests on statistical foundations which are clearly mathematical. Many exciting developments in mathematical statistics – including log-linear models, regression methods, diagnostics and treatments, structural equation modeling, and time series – have dramatically changed the ways of data analysts. Discussions of these procedures in sociological publications such as *Sociological Methodology* tend to be heavily mathematical. So much so that its audience has to be mathematically literate. Also, these statistical procedures are implemented both in main frame packages (such as SAS, BMDP, SPSSX) or in micro-computer packages (such as SYSTAT or STATA). If these tools are mobilized in the use of coherent models, there is little to which an objection can be made. They are powerful and efficient and, as a result, very useful and highly desirable. However, most of the time it appears that they are not used to estimate the parameters of a coherent model but for the assessment of some *ad hoc* construction.

There appear to be two broad reasons for this common practice. The first can be traced back to Zetterberg (1955). In essence, he attempted to establish formal reasoning from a set of axioms but without a mathematical model. As Fararo (1984:370) puts it, "formal theory, in this sense, was notational translation of verbal theories rather than a creative construction of testable idealized models based on exact concepts". Verbal statements were strung together in some kind of quasi syllogism. As Freeman (1984:345) notes, the problem of alleged logical reasoning with imprecise language is that the structure of the argument gets mixed up with the substance.

The second foundation on which contemporary sociological data analysis is founded is that of "causal modeling". Simon (1954) pointed out that, in a world modeled by linear relations, an empirical correlation between two variables could be seen as spurious by means of using partial correlations. Thus, if a correlation is noted between the number of fire trucks turning up at a fire and the damage done by the fire, the conclusion that fire trucks are dangerous things is not warranted. Controlling for the size of the fire (large fires do great damage and are fought with many fire trucks), the partial correlation between number of fire trucks and amount of damage is indistinguishable from zero. This led to the use of partial correlations, more or less as a geiger counter, to find the causal structure among a set of variables. Moreover, the claim was made that the technique could be used for non-experimental data (Blalock, 1963). Sifting correlations and partial correlations was replaced by path analysis (Wright, 1934, Duncan, 1966) where simple linear equations were strung together in a single model. In turn, path analysis evolved into fully fledged structural equation modeling based on the stunning work of Joreskog (1973). Either from verbal axiomatics or "causal modeling", a predominant concern is the estimation and interpretation of *linear* models. In one sense, the transition from asking "is there a relation among variables?" to "what is the form of the relation?" is a step in the right direction. But the relation is not one that is established by any coherent modeling procedure, mathematical or otherwise, it is simply asserted in making linear model techniques of immediate relevance but dubious utility. While running regressions and the like is easy, this whole effort may be doomed to failure precisely because it has no coherent theoretical process model alongside the quantified measurement.

Consider Kepler's struggle to show that each planet appeared to describe an ellipse having the sun as one of its foci. The relevant variables are the mean distance, r of its elliptical orbit from the sun, and the planetary year, t, the time it takes to complete a full orbit. After extended and amazing computations, given the tools he had, he found his third law: $t^2 = Kr^3$, with K a constant. Would a contemporary quantitatively inclined sociologist find it? It seems very unlikely given current practice. Consider the following "data" taken from Schiffer and Bowden (1984:83) for t and r.

t	r
287,496	484
1,601,613	1,521
2,146,689	1,849
4,251,528	2,916
4,721,632	3,136
7,414,875	4,225
9,261,000	4,900

The scattergram is shown in Fig. 2 where t is plotted against r.

One immediate act of our contemporary social scientists would be to compute the correlation between t and r or fit the linear relation between the two. The correlation would be computed as 0.99 and the amount of variance explained would be 98%. This would be one of the most staggering linear fits ever and the effort might well rest at this spot. Eyeballing Fig. 2 suggests some (slight) curvilinearity in the plot and, in an effort to capture this the most likely action is to regress t on r and r^2. This would give the variance explained as 99.99%. "O Frabjous day! callooh! callay!" If the effort was made to regress log t on r about 89% of the variance would be explained and the king of the heap would be to regress log t on log r, explaining all of the variance (in log r). With a data dredging establishment

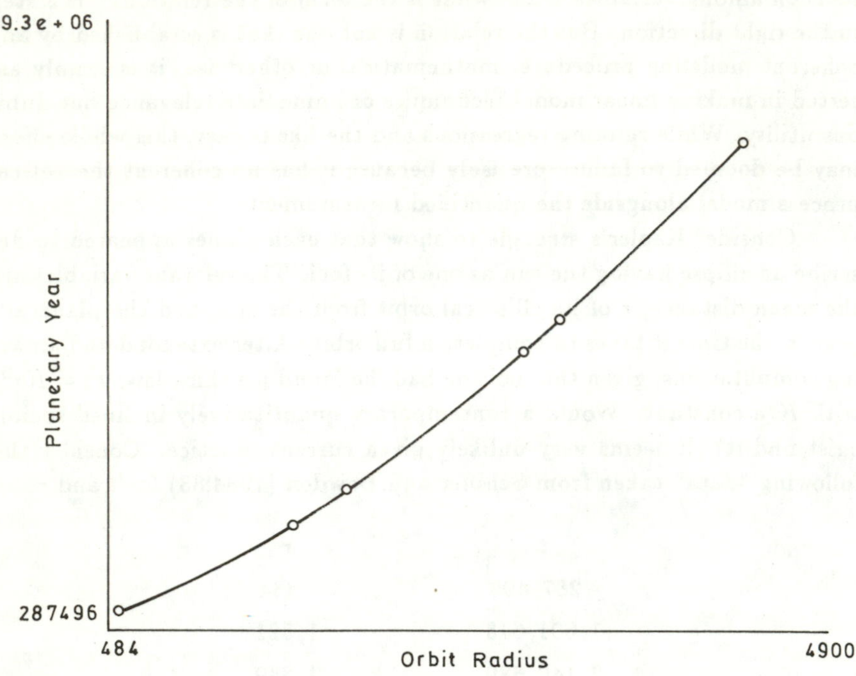

Fig. 2. Plot of planetary year against orbital radius.

of $\ln(t) = 1.47 + 3.30 \ln(r)$ it is highly unlikely that Kepler's third law would be "detected" in the data. Blind curve fitting of a simple functional form unmotivated by theory appears to head in the wrong direction. Of course, there may be no such laws for sociologists to find or establish. If there are, proceeding via coherent process models in mathematical form would be the way to go. This is, as noted, a minority pursuit.

4. Reasons for the Relative Absence of Mathematics in Sociology

There are many successful uses of mathematics within sociology. Leik and Meeker (1975) discuss in some detail a variety of models that have been used in mathematical sociology. Similarly, Rapoport (1983) discusses many models used in the social and behavioral sciences. Kim and Roush (1980) take it as a given that mathematics is useful, and advocate set theory and binary relations, matrix algebra, Boolean matrices, graph theory, combinatorics, difference equations, differential equations, and probability as viable modeling tools for studying social phenomena. There have been instances where a common mathematical formalism can model diverse social phenomena: the logistic model, for example, can model the growth of populations, the spread of prescribing a specific drug among physicians, the spread of infectious diseases, and consumer demand for products. Also in parallel to the natural sciences, Fararo and Skvoretz (1986) synthesize structural theory and expectations states theory in a single formal framework. So why the relative absence?

4.1. Spurious Reasons

Given such successes, one argument for the relative absence of mathematics in sociology is that most sociologists are not trained in mathematical social science or even appreciate it. Mathematical sociology is offered as a topic at relatively few places and, consistent with its marginal subspeciality standing, it seldom dominates the curriculum. However, the argument is trivially true and leaves unanswered the broader question as to why the majority of the sociological tribe ignore it. In a similar vein, it is quite unacceptable to argue that if more people knew the mathematics then the utility of it would be far more apparent.

Another reason could be the immense fragmentation of mathematical sociology. In part, the empirical problems studied are diverse and given the fragmentation that is characteristic of sociology, it is scarcely surprising that mathematical sociology is also fragmented. Hayes (1984:326) indicts "the failure to produce an integrated core of methods and results". Coleman (1964) even remarks on the battery of mathematical languages available, in principle, to the social scientist. The Rapoport (1983) and Leik and Meeker (1975) texts reflect the fragmentation that Hayes sees. But the fragmentation argument does not hold water either. Even if sociology will forever be fragmented, it could have a mathematical foundation.

Another completely unacceptable argument is that social phenomena are too complicated. Arguments like "God gave the easy problems to the physicists" do not bear scrutiny. A physical scientist setting out to observe and measure the trajectory of every snowflake in a blizzard, or every leaf as it falls from a tree in the autumn, or fragments of a house destroyed in a tornado would be regarded as rather foolish. While the physical principles underlying such motion are simple, the phenomena they generate are extremely complicated. To inductively establish the principles from such massive observation appears foolhardy. Yet, it seems that the social sciences are trying to do precisely this when great amounts of measurement are done – with or without a coherent theoretical perspective – with a view of representing all of the "richness and texture of social life" in some summary description, regardless of whether it is verbal, formal, or even poetic. By contrast, it appears that the natural sciences selected out a strategically chosen set of primitives/terms – mass, velocity, force, and pressure etc. Sociologists have yet to do this.

The "complexity of social life" is also reflected in the difficulty of measuring relevant variables. Concepts like anomie and alienation are hard to conceptualize, measure, and understand. But so too was "force" and it is not at all clear that the "gravitational pull" of Newton's theory is understood (Kline, 1985). The headlong rush to deal with complexity may also have moved sociology away from the true nature of theory construction and experimentation (Willer, 1984). The difficulties of measurement do not provide a sufficient excuse for the limited role of mathematics in sociology.

Also unacceptable is the idea that we need more time – a defense also proffered by most sociologists to account for the paucity of their results (Mayhew, 1980). Even three decades from the post World War II heady days seems long enough, but mathematical social science can be traced much further back to at least Quetelet in 1835 and his proposal for a social physics – although that seems more naturally to have evolved into statistical analysis (Stigler, 1986). We can go further back to Condorcet working in 1785 on the intransitive outcome of majority voting. Even if sociology did not start until 1855 with Comte, and the idea of using mathematics in social science was seriously entertained – so much so that it was seen as becoming the positive cap for the whole analytic hierarchy with mathematics at its foundation – it remains ludicrous to plead for more time.

4.2. Non-Spurious Reasons

Newton stands as a towering figure: not so much a pygmy on the shoulders of giants but a giant on the shoulders of giants. He produced the Newtonian revolution in science and set forth a new way of using mathematics in natural philosophy (Cohen, 1985). He invented calculus (as did Leibnitz) in which he could state his famous laws: the principle of inertia, the relation of force to its dynamic effect, and the equality of action and reaction. Hayes (1984) is quite devastating when he compares the statement of Newton's laws with a, seemingly identical, statement of Simon (1957) recasting Homans' (1950) hypotheses for small group behavior. While Newton's laws state a theory definitively, the social science parallel is little more than an explication of some of its elements. It is discomforting to know that Simon's analysis is still seen as one of the success stories in the applications of mathematics in sociology.

With regard to Newton, the need for the new mathematics stemmed directly from the physics of the problems he tackled. Other instances can be added – although mathematics can also be created for its sake and only later be found of direct relevance in science, usually in more than one area (for example group theory, Galois field theory, and matrix algebra).

To others, it looked as if "physicists adopted mathematical formalisms and physicists flourished" (Freeman, 1984:344). And so the impulse to imitate: if it worked there, maybe it can work here. Sørenson (1978) notes "most contributions to the literature apply mathematical models already existing and usually borrowed from other disciplines to sociological phenomena". As noted by Freeman (1984), the virtues of mathematics lie in the abstract form of the argument and that the structure of its reasoning is portable. However, the borrowing does not guarantee success, as critics of the use of mathematics in sociology are quick to point out.

We know that science is characterized by objectivity, generality and precision and that these properties are most clearly found in mathematics. It is not surprising that parsimony is served by mathematics. Nor is it surprising that specific formal models crop up in diverse contexts, even in social science. What seems lacking in sociology is the right/appropriate mathematics. The chances of success increase dramatically when any field "can either develop or cause to be developed mathematical tools that are designed specifically to map its peculiarities" (Freeman, 1984:347). Which puts us in a position of waiting for our Newton. Given the fundamental geometric role of Newton's mathematics, it is reasonable to hope for new geometries or envision alternative geometric representations. According to Willer (1984:243) "while Galileo had at least Euclidean geometry which could be reinterpreted for his uses, no equivalent geometry for social phenomena was present for use by the classical theorists. In fact, the opportunity to systematically generate theoretical models in sociology has required first the development of graph and network theory". Unfortunately for this example, graph theory was invented in 1736, with Euler solving the "Konigsberg problem" (Harary, 1969). Graph theory was around by the time of the classical theorists, but perhaps by then the potential connection with mathematic formulations had already dissolved. Willer is correct, however, in noting the need for some geometry or its equivalent. Other potential formal foundations, including algebraic topology and category theory, are considered in the next section. It is clear that there is a need for the conjunction of an appropriate mathematics and sociological problems.

5. Chances for Change

5.1. *Specialty Fields*

Every textbook presenting "mathematical ____________" (sociology, ecology, biology, psychology, economics or whatever) outlines the benefits of using mathematics. There clearly are many, but equally clearly, at least for sociology, the benefits are not persuasive for the larger tribe. It is evident also that exhortations and declarations are of limited value, at best. There are sociological theories, common methods, and models, the last of perhaps an unknown logical standing. And theories, methods, models, and practices are not replaced by critiques of them – only by better theories, better methods, better models and practices. Only successful applications of sufficient and compelling power as Newton's *Principia* will do. The time for a promissory note concerning the ultimate success of mathematical sociology is long passed – it is, after all, close to midnight. As noted already, for mathematics to be used on a wide scale in sociology requires major change. Even then it may be that the role of mathematics will not be fundamental. Wilson (1984:231–232) sees that "mathematical models have an essential place in our efforts to untangle the complexity of social reality". The essential role however is as a heuristic device rather than a fundamental tool. In a similar vein, Rapoport (1983) sees the predominant role of mathematical models as providing insight. Even with the role limited either to a heuristic strategy or an insight generator, mathematics would require a much larger role to be taken seriously as important for sociology. At issue then is how this change could come about.

To assess the chance of change requires that we look at how knowledge is developed, or more precisely, the social structure of the research community producing the knowledge as it impacts that knowledge. Price (1965) points out that scientific work goes on in research specialties when new knowledge models and methods are selectively attached to the old. Narin *et al.* (1973) document a mosaic of science as a network of subnetworks of ideas, etc. The differential density within these networks roughly coincides with the different disciplines.

The social relations of science and science specialties are not static and life-cycle models have been proposed to account for change through time (Crane, 1972, Mulkay *et al.*, 1975, and Mullins, 1973). Mullins outlines four stages of a life-cycle of an important specialty in science: normal, network, cluster, and specialty (or discipline). According to Mullins, the normal stage is characterized by limited organization, both in the literature and socially. He sees little coordinated effort in solving particular problems, but there will be intellectual leaders working on promising problems that produce successful research. Obviously, production and publication continue. In the network stage, social organizational leaders begin to emerge. These leaders generate one or more exciting intellectual products that interest others. The new intellectual leaders are able to recruit like-minded scientists and to attract students at specific locations. Within these groups, communication rises, the overall network of science becomes more differentiated, and programmatic statements emerge concerning the groups and its work. Next, the cluster stage is characterized by clusters of scientists and students forming around key intellectual leaders, again at a limited number of institutions, with highly inbred communication. Members of the clusters mutually reinforce each other's work and they concentrate on a specific set of highly important problems (at least to them) defined in their programmatic statements. During this process, secondary materials and critical assessments also emerge. To some extent, the in-group's research program becomes received dogma. For Mullins, the specialty stage is the last phase of the life-cycle. Students of, and other colleagues of, the leaders become successful themselves. In turn, they become geographically mobile as they pursue their scientific careers. The internal communication becomes less ingrown, and if successful, the cluster institutionalizes the work it has done. The period of routinization that follows sets up the next stage of normal science.

Mullins applies his model to a variety of theory groups in contemporary American sociology. Two of these groups are particularly relevant for this discussion. The first is the causal modeling group described briefly in the section on quantification. The mathematics implicitly used by this "new causal theory" group has made the largest inroads into institutionalized sociology, at least in the United States and in professional associations. But by prematurely clinging to an overly simplified guess as to a "formal" representation there is real doubt that this group is "causal" or even that it is the legitimate heir of the positivist tradition. It is heir to the technological revolution inherent in computers and the invention of the social survey. The ability to conduct social surveys vastly increased the ways in which sociologists can get a reading on social phenomena. The availability of a computer made the analysis of vast quantities of data possible. There is a great irony in these developments. Social scientists in general, and sociologists in particular, have traditionally been concerned with the analysis of social structure and the context of that structure. Sample surveys, while technically convenient, pull individuals, organizations, and institutions out of their structural context destroying perhaps the most important part of social reality. This provocative claim has become the received dogma of another theory group that Mullins discusses.

It is to the "structuralists", or as they are now known the "social network analysts", that we next turn. This group is heir to another technological development in the form of readily available specially written software and microcomputers that can be used to analyze social structure. Moreover, this group is over fifteen years older than when they were studied by Mullins.

Many students and social scientists have been attracted to network analysis and it is rightly viewed as a dynamic specialty. It has passed through all four of the stages delineated by Mullins and it satisfies the first two of Cohen's criteria as to whether or not a scientific revolution has taken place. It is probably too early to assess the third and fourth, but some would claim it passes the third criterion also.

Of interest here is the fact that many of the social network analyses are explicitly mathematical in their orientation. According to Freeman (1984) the study of social networks is nothing if not mathematical and, from the 1960s onwards, graph theory has been an often mobilized tool. This group has its own professional association (the *International Network of Social Network Analysts*), its own journal (*Social Networks*) and a clearly defined research agenda. The journal, *Social Networks*, has been described as inducing "symbol shock", a sure indicator that its language is not routine. With respect to its mathematical orientation, mathematicians have been involved in the group and many network analysts have a strong training in mathematics. Rightly or wrongly, they have nailed their flag to the mathematics flagpole and will succeed or fail with its mathematical core. Already, there are people interested in network analysis who think that the use of mathematics has gone too far (for example, Boissevain, 1979; Rogers and Kincaid, 1981). However, it is more likely that Barnes and Harary (1983) are closer to the mark when they say that, quite to the contrary, the group has not used enough of the powerful mathematics available within graph theory.

Social network analysts tend to define themselves as a distinct specialty concerned with structural analysis. Among its practitioners are sociologists, anthropologists, political scientists, social psychologists, mathematicians, communication theorists, and information scientists. For some of the sociologists, including me, the disciplinary home has been abandoned even if they continue to tackle profoundly sociological issues. Hayes (1984) gently chides me for having written "the future of sociology as a viable discipline will largely depend on the use of mathematics in an informed and imaginative manner" (Doreian, 1970). Whilst I cleave to that view, the optimism with regarding sociology is dimmed, and it may not be a viable scientific discipline.

5.2. Structural Mathematics

In the mid-1960's, Harary, Cartwright, and Norman (1965) published a book entitled *Structural Models.* It was an extended development of directed graph theory which became an invaluable tool to those representing social networks. Two decades later, Hage and Harary (1986) published *Structural Models in Anthropology*, discussing many ways in which directed graphs had been used in that field. Category theory was used by Lorrain and White (1971) as the mathematical realm within which the fundamental properties including equivalence in social networks could be founded. For them, the analysis of social structure proceeded by a careful examination of the objects and morphisms in a category. As it turned out, the computational problems in such an enterprise were fearsome and the structuralists may have tried a shortcut, much like the one taken by the curve fitters and causal modelers. Some quick and dirty algorithms have been constructed that roughly provide measures of structural equivalence.

Structural equivalence has been generalized, mathematically, by the concept of regular equivalence (White and Reitz, 1983) and, more recently, Everett, Boyd, and Bortgatti (1988) have proposed orbit equivalence. Doreian (1988) has pointed to the real need to ensure that the algorithms used within network analysis are faithful to the mathematical and social conceptions on which they are based. There is the serious risk that, in Willer's terms, that network analysts will jump directly from action to structure in a way that is as crude as the premature specification of linear relationships by the new causal theorists.

Algebraic topology in the form of Q-analysis, may prove another valued tool in the social network analyst's tool kit and was developed by Atkin (1977, 1981) to deal with the specific analysis of social phenomena. The major stumbling block in the path of its use is a very clear requirement that sets be clearly and accurately defined. Given social scientists' proclivity for murky and ill-defined terms, this is a stringent requirement but, as Atkin and others have claimed, it is only by getting such – qualitative – measurement right in the first place that any intelligible results can be generated. By defining a structural backcloth, the Q-analysts do have a geometry and distributed changes over that geometry can be interpreted in an analogous fashion to the treatment of force in Euclidean space. Their deep intuition is that social phenomena will occur only when this structural backcloth permits it. There have been relatively few attempts to mobilize this machinery in sociological context (for example Doreian (1982), Freeman (1981)). The major point here is that the mathematics is being developed in response to sociological problems.

There are many areas in which nonquantitative methods are being applied to structural problems. The duality of people and groups has been elegantly exploited by Breiger (1974) with the use of bipartite graphs. For sociological reasons, his analysis has been extended by Fararo and Doreian (1986) to tripartite graphs. With regard to algebra, group theory and semi-group theory have been mobilized in the algebraic analysis of social relations. This work founded in the kinship analysis of White (1963), in an effort to grapple with the conceptual subtleties of social structure (Nadel, 1957), evolved into the structural equivalence ideas noted earlier. But remaining in the algebraic model, rather than jumping immediately back to an empirical interpretation, has created the potential for generating an algebra of social relations. Among the exciting developments in this area is the analysis of cumulated social roles of Breiger and Pattison (1986).

This overly brief listing of the use of some mathematical ideas in social network analysis points to a bubbling, a ferment, of ideas that may, or may not, provide the basis for the creation of a science of society. It is a fundamentally different kind of mathematical analysis than that embraced by the "new causal theorists" and the kinds of theories and constructions are far more in keeping with the style of theory construction outlined by

Fararo (1984).

6. Conclusion

Science is full of dead-ends and false trails. There is no reason to expect that sociology, with its warring factions, will be any different. Indeed, because it lacks disciplinary coherence in its pre-paradigm state, it is better characterized as social studies rather than social science. Despite Comte's forward looking view concerning the future of science and sociology as a positive science, the discipline is very far removed from fulfilling his vision. While there have been a series of "small" successful applications of limited models to specific social phenomena, there is no fundamental mathematical foundation for sociology. Even within economics, where mathematization is taken far more seriously, the mathematical economists concede that their elegant models have no predictive value. Yet the hallmark of successful application in the physical sciences is incredibly accurate prediction – even with inaccurate measurement. A debate as to whether or not the mathematical structure is "out there" in the physical universe waiting to be discovered is reasonable, even though it may be undecidable. Within the social sciences there is no basis for such a debate as the successes are minuscule.

Although some social scientists have imitated the natural sciences and have used or borrowed mathematics created for other purposes, it seems that only the creation of, and mobilization of, mathematics specifically attuned to the sociological concerns can provide a fundamental foundation. For this to occur would take a far reaching synthesizing revolution in sociology.

Individual sociologists make their own commitments and only a small handful pursue the mathematical vision. No longer can they promise a mathematical future for sociology. "In principle", debates are less than useful, devouring needed time and energies. All that really counts by way of convincing evidence are convincing demonstrations. I have argued that in social network analysis there are the stirring of what might become a genuine revolution. But it is only stirring and individuals make use of whatever mathematics they can. This group will pursue its dream and only when they succeed, or are shown to have failed miserably, will there be any convincing evidence from the subfield. In the marketplace of ideas, truth will out in the long run and only in the long run will we have an inkling of the true value, if any, of mathematics in sociology – or in the social sciences.

There are other promising lines of inquiry, rooted in mathematics, that have not been covered here. Hannan's work on event – history analysis (for a summary see Hannan (1988)) and the whole area of stochastic model building (see Bartholomew (1982)) contain models and methods that can be foundational. The majority of sociologists will continue their work without mathematics and the arguments over the best or most fruitful approach will continue. Given the heat of prior debates, this will also be a tale of sound and fury. My hope is that it will not be a tale told by idiots nor that it signifies nothing, and that we have a mathematical carriage rather than a pumpkin.

References

1. R. H. Atkin, *Combinatorial Connectivities in Social Systems* (Birkhauser Verlage, 1977).
2. R. Atkin, *Multidimensional Man* (Penguin, 1981).
3. J. A. Barnes and F. Harary, *Graph Theory in Network Analysis, Social Networks* **5** (1983) 235–244.
4. D. J. Bartholomew, *Stochastic Models of Social Processes*, 2nd edition (Wiley, 1982).
5. H. M. Blalock, *Causal Inferences in Nonexperimental Research* (Univ. of North Carolina Press, 1964).
6. J. Boissevain, *Network Analysis: A reappraisal, Current Anthropology* **20** (1979) 392–394.
7. T. Bottomore and R. Nisbet, *A History of Sociological Analysis* (Free Press, 1978).
8. R. L. Breiger, *The Duality of Persons and Groups, Social Forces* **53** (1974) 181–190.
9. R. L. Breiger and P. E. Pattison, *Cumulated Social Roles: The Duality of Persons and Their Algebras, Social Networks* **8** (1986) 215–256.
10. I. D. Cohen, *Revolution in Science* (Belknap Press, 1985).
11. J. S. Coleman, *An Introduction to Mathematical Sociology* (Free Press, 1964).
12. A. Comte, *The Positive Philosophy of August Comte* (*Calvin Blanchard*, 1855), translator Harriet Martineau.
13. D. Crane, *Invisible Colleges: Diffusion of Knowledge in Scientific Communities* (Univ. of Chicago Press, 1972).
14. O. D. Duncan, *Path Analysis: Sociological Examples, Am. J. of Sociology* **72** (1966) 1–16.
15. P. Doreian, *Mathematics and the Study of Social Relations* (Weidenfeld and Nicolson, 1970).
16. P. Doreian, *Leveling Coalitions as Network Phenomena, Social Networks* **4** (1982) 25–45.
17. P. Doreian, *Equivalence in a Social Network, J. Math. Sociology* **13** (1988) 243–282.
18. M. Everett, J. Boyd and S. Borgatti, *Ego-centered and Local Roles: A Graph Theoretic Approach, J. Math. Sociology* **15** (1989) forthcoming.
19. T. J. Fararo, *Mathematical Sociology* (John Wiley, 1973).

20. T. J. Fararo, *Neoclassical Theorizing and Formalization in Sociology, J. Math. Sociology* **10** (1984) 361–393.
21. T. J. Fararo and P. Doreian, *Tripartite Structural Analysis: Generalizing The Breiger-Wilson Formalizm, Social Networks* **6** (1984) 141–175.
22. T. J. Fararo and J. Skvoretz, *E-State Structuralism: A Theoretical Method, Am. Sociological Rev.* **15** (1986) 591–602.
23. L. C. Freeman, *Q-Analysis and the Structure of Friendship Networks, Int. J. Man-Machine Studies* **12** (1981)
24. L. C. Freeman, *Turning a Profit from Mathematics: the case of Mathematics, J. Math. Soc.* **10** (1984) 343–360.
25. R. W. Hamming, *The Unreasonable Effectiveness of Mathematics, Am. Math. Monthly* **87** (1980) 81–90.
26. M. T. Hannan, *Age Dependence in the Mortality of National Labor Unions: Comparisons of Parametric Models, J. Math. Sociology* **14** (1988) 1–30.
27. F. Harary, *Graph Theory* (Adison-Wesley, 1969).
28. F. Harary, R. Z. Norman and D. Cartwright, *Structural Models: An Introduction to the Theory of Directed Graphs* (John Wiley, 1965).
29. A. C. Hayes, *Formal Model Building and Theorectical Interest in Sociology, J. Math. Sociology* **10** (1984) 325–341.
30. D. D. Heckathorn, *Mathematical Theory Construction in Sociology: Analytic Power, Scope, and Descriptive Accuracy as Tradeoffs, J. Math. Sociology* **10** (1984) 295–323.
31. G. C. Homans, *The Human Group* (Harcourt, Brace and World, 1950).
32. K. G. Joreskog, *A General Method for Estimating a Linear Structural Equation System, Structural Equation Models in the Social Sciences*, eds. A. S. Goldberger and O. D. Duncan (Seminar Press, 1973).
33. K. H. Kim and F. W. Roush, *Mathematics for Social Scientists* (Elsevier, 1980).
34. M. Kline, *Mathematics and the Search for Knowledge* (Oxford University Press, 1985).
35. T. S. Kuhn, *The Structure of Scientific Revolutions*, 2nd edition (Chicago Univ. Press, 1977).
36. C. A. Lave and J. G. March, *An Introduction to Models in the Social Sciences* (Harper and Row, 1975).
37. R. F. Leik and B. F. Meeker, *Mathematical Sociology* (Prentice-Hall, 1975).
38. G. Lenski and J. Lenski, *Human Societies: An Introduction to Macro Sociology*, 3rd edition (McGraw Hill, 1978).
39. F. Lorrain and H. C. White, *Structural Equivalence of Individuals in Social Networks, J. Math. Sociology* **1** (1971) 49–80.
40. B. H. Mayhew, *Structuralism Versus Individualism: Part I, Shadowboxing in the Dark, Social Forces* **59** (1980) 335–375.
41. M. J. Mulkay, G. N. Gilbert and S. Woolgar, *Problem Areas and Research Networks in Science, Sociology* **9** (1975) 187–203.
42. N. C. Mullins, *Theories and Theory Groups in Contemporary American Sociology* (Harper & Row, 1973).

43. S. F. Nadel, *The Theory of Social Structure* (Cohen and West, 1957).

44. F. Narin, M. Carpenter and N. C. Berlt, *Interrelationships of Scientific Journals, J. Am. Infor. Sci.* **22** (1972) 323–331.

45. M. Olinick, *An Introduction to Mathematical Models in the Social and Life Sciences* (Addison Wesley, 1978).

46. D. J. de S. Price, *Networks of Scientific Papers, Science* **149** (1965) 510–515.

47. A. Rapoport, *Mathematical Models in the Social and Behavioral Sciences* (John Wiley, 1983).

48. F. S. Roberts, *Discrete Mathematical Models* (Prentice-Hall, 1976).

49. D. R. White and K. P. Reitz, *Graph and Semigroups Homomorphisms on Networks of Relations, Social Networks* **5** (1983) 193–234.

50. E. M. Rogers and D. L. Kincaid, *Communication Networks: Towards a Paradigm for Research* (Free Press, 1981).

51. M. M. Schiffer and L. Bowden, *The Role of Mathematics in Science* (Mathematical Association of America, 1984).

52. H. A. Simon, *Spurious Correlation: A Causal Interpretation, J. Am. Statistical Assoc.* **49** (1954) 467–479.

53. H. A. Simon, *Models of Man* (John Wiley, 1957).

54. *Sociology: An Introduction*, ed. *N. J. Smelser* (John Wiley, 1967).

55. A. B. Sørensen, *Mathematical Models in Sociology, Ann. Rev. Sociology* **4** (1978) 3345–71.

56. H. C. White, *The Anatomy of Kinship* (Prentice-Hall, 1963).

57. S. M. Stigler, *The History of Statistics: The Measurement of Uncertainty Before 1900* (The Belknap Press, 1986).

58. E. P. Wigner, *The Unreasonable Effectiveness of Mathematics in the Natural Sciences, Commun. Pure and Appl. Math.* **13** (1960) 1–14.

59. T. P. Wilson, *On the Role of Mathematics in the Social Sciences, J. Math. Sociology* **10** (1984) 221–239.

60. S. Wright, *The Method of Path Coefficients, Ann. Math. Stat.* **5** (1934) 161–215.

61. H. L. Zetterberg, *On Theory and Verification in Sociology* (Almquist and Wiksell, 1954).

Fundamental Roles of Mathematics in Science

Donald Greenspan
Department of Mathematics
University of Texas at Arlington
Arlington, TX 76019, USA

1. What is Mathematics?

Mathematics is a collection of disciplines with names like Euclidean geometry, Riemannian geometry, algebra, group theory, loop theory, ring theory, calculus, real analysis, complex analysis, functional analysis, operator theory, differential equations, integral equations, statistics, and probability. Unifying all these disciplines are *four criteria* which each special area of mathematics *must* satisfy. These criteria are described in simple terms as follows.

Let us begin, and, perhaps, seemingly quite apart from our subject, by considering some of the extant problems associated with communication between people by means of language. If any particular word, like *ship*, were flashed on a screen before a large audience, it is doubtful that any two people would form exactly the same mental image of a ship. It follows, similarly, that the meaning of every word is so intimately related to a person's individual experiences, that probably *no word has exactly the same meaning to any two people.*

To further complicate matters, *it does not appear to be possible for anyone to ever find out what a particular word means to anyone else.* Suppose, for example, that man X asks man Y what the word *ship* means to him and Y replies that a *ship* is a vessel which moves in, on, or under water. X, realizing that even a rowboat tied to a pier is moving by virtue of the earth's rotation, asks Y to clarify his definition of *ship* by further defining *to move.* Y replies that *to move* is to relocate from one position to another by such processes as walking, running, driving, flying, sailing, and the like. X, for exactness, then asks Y if by *sailing* he means the process of navigating a ship which has sails, to which Y replies yes. "Then", replies X, "I shall never be able to understand you. You have defined *ship* in terms of *move, move* in terms of *sailing*, and *sailing* in terms of *ship*, which was the word originally requiring clarification. You have simply talked around a circle."

The circular process in which X and Y became involved so quickly is indeed one in which we can all become entangled if we constantly require definitions of words used in definitions. For the total number of words in all existing languages is finite and it would be merely a matter of time to complete a cycle of this verbal merry-go-round.

Now, in constructing the language of a mathematical discipline, the mathematician examines the two semantic problems described above and agrees that no two people may ever understand *completely* what any particular word means to the other. With this supposition, however, the problem of definitions resulting in a circular process can be, and is, avoided as follows. Suppose, says the mathematician, the words

a	in	path
by	is	point
direction	move	the
fixed	out	trace

are called *basic terms* and are stated without definition. We all have ideas and feelings about these words, but rather than attempt to make their meanings precise to each other, we shall simply leave them undefined. Now, let us define a *line* as the path traced out by a point moving in a fixed direction. Note that the word *line* is defined only in terms of the basic terms. Next, define a *plane* as the path traced out by a line moving in a fixed direction. Note that *plane* is defined in terms of only *line* and of basic terms. Now suppose that X asks mathematician Y what a *plane* is. Y responds that a *plane* is the path traced out by a line moving in a fixed direction. X, for clarity, asks Y what he means precisely by a *line*, to which Y responds that a *line* is the path traced out by a point moving in a fixed direction. X, seeking further clarity, asks for the definition of *point*, to which the mathematician responds, "*Point* is an undefined basic term," and there the questioning stops.

Thus, the first criterion is that *every mathematical discipline begins with basic terms which are undefined and all other concepts are defined by means of these. Point* is an undefined concept of geometry and *positive integer* is an undefined concept of algebra. No other subject treats its notions this way.

But let us look a bit further into the nature of mathematical concepts. Consider, for example, the geometric concept called a *straight line*. With a pencil and ruler, we have all at one time or another drawn a straight line. But, indeed, have we really ever drawn a straight line? A mathematical line has *no* width, while the line we draw with pencil and ruler certainly does have some width, even though one might need a special instrument, like a micrometer, to measure the width. As a matter of fact, the width may even vary as the pencil lead is being used up in the drawing process.

Indeed, *every* physical object has some width and it must follow that the mathematical straight line is an idealized form which exists only in the mind, that is, it is an abstracton. In a similar fashion, it can be shown that *all mathematical concepts are idealized forms which exist only in the mind,* that is, are abstractions. This is the second criterion for a mathematical discipline.

So, all mathematical concepts are abstractions which either are undefined or have definitions constructed on basic undefined terms.

After having constructed a system of concepts, the mathematician next seeks a body of rules by which to combine and manipulate his concepts. Thus, mathematical disciplines now take on the aspects of a game in that the rules of play, which must be followed, have to be enumerated. *Each mathematical discipline has its own rules of play or what are technically called assumptions or axioms,* and this is the third criterion. The axioms of algebra are, indeed, quite simple. For example, for the numbers 2, 3 and 5, it is assumed that

$$2+5=5+2\,,$$
$$2\times 5=5\times 2\,,$$
$$(2+3)+5=2+(3+5)\,,$$
$$(2\times 3)\times 5=2\times(3\times 5)\,,$$
$$2\times(3+5)=2\times 3+2\times 5\,.$$

In complete abstract form, then, if a, b and c are three positive integers, the algebraist assumes that

$$a+b=b+a\,, \qquad \text{(Commutative axiom of addition)}$$
$$a\times b=b\times a\,, \qquad \text{(Commutative axiom of multiplication)}$$
$$(a+b)+c=a+(b+c)\,, \qquad \text{(Associative axiom of addition)}$$
$$(a\times b)\times c=a\times(b\times c)\,, \qquad \text{(Associative axiom of multiplication)}$$
$$a\times(b+c)=a\times b+a\times c\,, \qquad \text{(Distributive axiom)}\,.$$

The question which immediately presents itself is: How does one go about selecting axioms? Historically, axioms were supposed to coincide with fundamental physical concepts of truth. But, as the nineteenth century chemists and physicists began to destroy the previous century's physical truths, the choice of mathematical axioms became a relatively free one. And, indeed, it is a rather simple matter to show that the axioms stated above for numbers can be false when applied to physical quantities. For example, if a represents sulphuric acid and b represents water, while $a + b$ represents adding sulphuric acid to water and $b + a$ represents adding water to sulphuric acid, then $a + b$ is not equal to $b + a$ because $b + a$ can result in an explosion whereas $a + b$ cannot.

In this connection the history associated with the fifth postulate of Euclid is of scientific significance. Indeed, in Euclid's Elements, set forth in about 300 B.C., plane geometry was founded on ten axioms, five of which were called common notions and five of which were called postulates. The axiom of interest to us is the fifth postulate, stated usually in the following equivalent form due to Playfair:

Postulate 5 Through a point not on a given line, one and only one parallel can be drawn to the given line.

Through the centuries, Postulate 5 was of serious concern to mathematicians. Euclid himself seems to have avoided its use in proofs whenever possible. The reason for its somewhat tenuous position among the other axioms of geometry lay in the realization that it was an assumption, and the only assumption, about an *infinite* object, that is, the entire straight line, when science knew of no physical object with any infinite quality or dimension. Indeed, even today, atomic and molecular theory maintain that everything in the material world is of finite character. But such a rock of Gibralter was geometry in the realm of mathematics, physics, and astronomy, that instead of seeking a physically acceptable replacement for Postulate 5, mathematicians until the nineteenth century sought primarily to establish its truth.

It was not until the latter part of the eighteenth century and during the nineteenth century that such men as Gauss, Bolyai, and Lobachevsky developed a second geometry in which Postulate 5 was replaced by the assumption that through a point not on a given line, one could draw *two* parallels to the given line. Still later, Riemann developed a third geometry by assuming that *no* parallels to a given line could be drawn through a point not on the line. And perhaps the greatest scientific impact of these new geometries was that the geometry of Riemann laid the groundwork for the geometry utilized in the Einstein's theory of relativity. Thus, mathematical history shows that until some freedom of choice with regard to the selection of axioms was realized, the development of the theory of relativity simply was not possible.

Note, however, that *complete* freedom of choice in the selection of axioms is available to no man. It would be impractical and nonsensical, for example, to start with assumptions like:

Axiom 1 All numbers are positive.

Axiom 2 Some numbers are negative.

for these assumptions contradict each other. Indeed, even though a set of axioms does not contain a contradictory pair, it can happen that reasoning from them would yield contradictory conclusions. But, further examination of the very deep problems associated with selection of axioms would be beyond the present scope.

The final difference between a mathematical discipline and a non-mathematical discipline lies in the reasoning processes allowed in reaching conclusions. There are basically two acceptable types of reasoning in scientific work, inductive reasoning and deductive reasoning. Let us consider each in turn.

Suppose scientist X injects 100 monkeys with virus Y and does not so inject a control group of 100 monkeys. One week later, ninety monkeys in the first group and only five in the second group contract flu. X, sensing a discovery, repeats the experiment and finds approximately the same statistical results. Further experiments are made in which various environmental factors like heat, light, proximity of cages, and so forth are varied, and in every case X finds that from 85% to 95% of the monkeys receiving virus Y become ill, while only from 3% to 10% the control group acquire the disease. X concludes that virus Y is the cause of flu in monkeys, and the process of reaching his conclusion by experimentation with control is called inductive reasoning. Note that if, after proving his result, X were to inject only one monkey with virus Y, all that he could say would be that the *probability* is very high that the monkey will become ill. Indeed, it is not absolutely necessary that flu will result.

Suppose now that mathematician X writes down a set of assumptions, two of which are

Axiom 1 All heavenly bodies are hollow.

Axiom 2 All moons are heavenly bodies.

Then it *must* follow, *without exception*, that

Conclusion: All moons are hollow.

The above type of reasoning from axioms to *necessary* conclusions is called deductive reasoning. The simple three line argument presented above is called a syllogism. The general process of reaching necessary conclusions from axioms is called deductive reasoning and the syllogism is the fundamental unit in all complex deductive arguments. In a mathematical discipline, *all conclusions* must be *reached by deductive reasoning alone*, which is the fourth and final criterion of a mathematical discipline. Although, very often *axioms* are selected after extensive inductive reasoning, no mathematical *conclusion* can be so reached. Thus, there is no question of a mathematical conclusion having a high probability of validity as in the case of inductive conclusions. Indeed, if the axioms are absolute truths, then so are the deductive conclusions.

Thus we see that each mathematical discipline deals with abstract idealized forms which are defined from basic undefined terms, relates its concepts by means of axioms, and establishes conclusions only by deductive reasoning from the axioms. It is the perfect precision of abstract forms and deductive reasoning which makes mathematics *exact*.

2. What is Science?

Let us turn now to characterizing science. Broadly speaking, science is the study of Nature in all its various forms. It appears that we study Nature not only because we are curious, but because we would like to control its very powerful forces. Understanding the ways in which Nature works might enable us to grow more food, to prevent normal cells from becoming cancerous, and to develop relatively inexpensive sources of energy. In cases where control may not be possible, we would like to be able to predict what will happen. Thus, being able to predict when and where an earthquake will strike might save many lives, even though, at present, we have no expectation of being able to prevent a quake itself.

The discovery of knowledge by scientific means is carried out in the following way. First, there are experimental scientists who, as meticulously as possible, reach conclusions from experiments and observations. Since no one is perfect, not even a scientist, and since experiments cannot be reproduced exactly, all experimental conclusions have some degree of error. Hopefully, the error will be small. Then, there are the theoretical scientists, who create models from which conclusions are reached. Experimental scientists are constantly checking these models by planning and carrying out new experiments. Theoreticians are constantly refining their models by incorporating new experimental results. The two groups work in a constant check-and-balance refinement process to create knowledge. And only after extensive experimental verification and widespread professional agreement is a scientific conclusion accepted as valid.

3. Fundamental Roles of Mathematics in Science

Finally, let us clarify the fundamental roles which mathematics plays in science. In general, scientific experimentation and observation require methodology for handling and analyzing data sets. Scientific modeling requires methodology for organizing a consistent theory and for solving equations and systems of equations. And, it is mathematics which provides constructive techniques for *both* these areas of endeavor.

More specifically, in the analysis of data sets, not only does the *proper* use of statistical and probabilistic methods enable one to determine the reliability and interpretation of data, but, for example, interpolation, extrapolation, and least squares techniques enable one to make reasonable predictions. For exceptionally large data sets, such methodology is now implemented on modern digital computers.

In formulating scientific theories, the *exactness* of a mathematical discipline is sought by organizing a system of assumptions, often, but not always, based on experimental results, and then by applying the formalism of deductive reasoning. At present, perhaps the most mathematical of the theoretical sciences are Newtonian mechanics, relativistic mechanics, and quantum mechanics, each of which is a physical science. One would, however, be hard put to isolate the undefined terms in each of these theoretical sciences because, too often, physicists are as careless about their mathematics as mathematicians are careless about their physics.

Finally, as for the importance of mathematics in formulating and solving equations, one cannot over-estimate the importance of having available mathematical concepts like derivatives, integrals, periodic functions, delta functions, and eigenfunctions, each of which has a direct physical analogue. As an example, consider the derivative. In science, one of the broad principles is that all things in Nature change with time. Some things change relatively slowly, like the erosion of rocks by wind and the movement of continents. Other things change relatively quickly, like the shape of wind-blown cloud and the color of a chamelion when it moves from brown earth into green grass. But, all things change with time. Now, the *derivative with respect to time* is that mathematical concept which describe *instantaneous rate of change with respect to time.* Hence, if one wishes to formulate a dynamical equation to describe how change with respect to time is occurring, that equation will have derivatives in it, and, at its simplest, will be a differential equation. Thus, differential equations are the fundamental equations of science, and it is no wonder that the fundamental dynamical equations of, for example, Newtonian mechanics, relativistic mechanics, and quantum mechanics are such equations. Great advances in theoretical science were made possible over the 300 year period from 1650–1950 because mathematics provided general, analytical methods for solving large classes of *linear* differential equations. With the development of modern computer technology and numerical methodology for approximating solutions of *nonlinear* differential equations, there has resulted an unprecedented explosion of knowledge in the physical sciences since 1950.

4. Misuses of Mathematics

Mathematical power, like other types of power, suffers too often from misuse. In much of modern quantum mechanics, for example, mathematical methods which yield *relative* minima are misused to reach conclusions which can only follow from knowledge about *absolute* minima. In general relativity, incorrect physical implications were deduced from the Schwartzhild solution of the field equations before it was realized that many other solutions existed, the problem being that no initial or boundary constraints were ever prescribed. In economics and psychology, *linear* least squares or *constant* input-output matrices are often used, and indeed, sometimes used automatically by means of canned programs, when fundamental force interactions are *nonlinear*. Unfortunately, linear models may be very poor approximations for nonlinear models.

The power of mathematics is in its precision. The precision of mathematics must be used precisely. When it is not so used, the quality of our scientific endeavors suffers.

Inner Vision, Outer Truth

Reuben Hersh
Department of Mathematics and Statistics
University of New Mexico
Albuquerque, NM 87131, USA

Here is an old conundrum, many times resurrected: why do mathematics and physics fit together so surprisingly well? There is a famous article by Eugene Wigner, or at least an article with a famous title: "The Unreasonable Effectiveness of Mathematics in Natural Sciences." After all, pure mathematics, as we all know, is created by fanatics sitting at their desks or scribbling on their blackboards. These wild men go where they please, led only by some notion of "beauty", "elegance", or "depth", which nobody can really explain. Wigner wrote, "It is difficult to avoid the impression that a miracle confronts us here, quite comparable in its striking nature to the miracle that the human can string a thousand arguments together without getting itself into contradictions, or to the two miracles of the existence of laws of nature and of the human mind's capacity to divine them."

In such examples as Lobachevsky's non-Euclidean geometry, or Cayley's matrix theory, or Galois' and Jordan's group theory, or the algebraic topology of the mid-twentieth century, pure mathematics seemed to have left far behind any physical interpretation or utility. And yet, in the cases mentioned here, and many others, physicists later found in these "useless" mathematical abstractions just the tools they needed.

Freeman Dyson writes, in his Foreword to Monastyrsky's *Riemann, Topology, and Physics*, of "one of the central themes of science, the mysterious power of mathematical concepts to prepare the ground for physical discoveries which could not have been foreseen or even imagined by the mathematicians who gave the concepts birth."

On page 135 of that book, there is a quote from C. Yang, co-author of the Yang-Mills equation of nuclear physics, speaking in 1979 at a symposium dedicated to the famous geometer, S. Chern.

"Around 1968 I realized that gauge fields, non-Abelian as well as Abelian ones, can be formulated in terms of nonintegrable phase factors, i.e., path-dependent group elements. I asked my colleague Jim Simons about the mathematical meaning of these nonintegrable phase factors, and he told me they are related to connections with fibre bundles. But I did not then appreciate that the fibre bundle was a deep mathematical concept. In 1975 I invited Jim Simons to give to the theoretical physicists at Stony Brook a series of lectures on differential forms and fibre bundles. I am grateful to him that he accepted the invitation and I was among the beneficiaries. Through these lectures T. T. Wu and I finally understood the concept of nontrivial bundles and the Chern-Weil theorem, and realized how beautiful and general the theorem is. We were thrilled to appreciate that the nontrivial bundle was exactly the concept with which to remove, in monopole theory, the string difficulty which had been bothersome for over forty years [that is, singular threads emanating from a Dirac monopole].... When I met Chern, I told him that I finally understood the beauty of the theory of fibre bundles and the elegant Chern-Weil theorem. I was struck that gauge fields, in particular, connections on fibre bundles, were studied by mathematicians without any appeal to physical realities. I added that it is mysterious and incomprehensible how you mathematicians would think this up out of nothing. To this Chern immediately objected. "No, no, this concept is not invented-it is natural and real."

Why does this happen?

Is there some arcane psychological principle by which the most original and creative mathematicians find interesting or attractive just those directions in which Nature herself wants to go? Such an answer might be merely explaining one mystery by means of a deeper mystery.

On the other hand, perhaps the "miracle" is an illusion. Perhaps for every bit of abstract purity that finds physical application, there are a dozen others that find no such application, but instead eventually die, disappear and are forgotten. This second explanation could even be checked out, by a doctoral candidate in the history of mathematics. I have not checked it myself. My gut feeling is that it is false. It seems somehow that most of the mainstream research in pure mathematics does eventually connect up with physical applications.

Here is a third explanation, a more philosophical one that relies on the very nature of mathematics and physics. Mathematics evolved from two sources, the study of numbers and the study of shape, or more briefly, from arithmetic and visual geometry. These two sources arose by abstraction or observation from the physical world. Since its origin is physical reality, mathematics can never escape from its inner identity with physical reality. Every so often, this inner identity pops out spectacularly when, for example, the geometry of fiber bundles is identified as the mathematics of the gauge field theory of elementary particle physics. This third explanation has a satisfying feeling of philosophical depth. It recalls Leibnitz's "windowless monads", the body and soul, which at the dawn of time God set forever in tune with each other. But this explanation, too, is not quite convincing. For it implies that all mathematical growth is predetermined, inevitable. Alas, we know that is not so. Not all mathematics enters the world with that stamp of inevitability. There is also "bad" mathematics, that is, pointless, ugly, or trivial. This sad fact forces us to admit that in the evolution of mathematics there is an element of human choice, or taste if you prefer. Thereby we return to the mystery we started with. What enables certain humans to choose better than they have any way of knowing?

A good rule in mathematical heuristics is to look at the extreme cases – when $\varepsilon \to 0$, or $n \to \infty$. Here, we are studying the way that discoveries in "pure" mathematics sometimes turn out to have important, unexpected uses in science (especially physics). I would like to use the same heuristic – "look at the extreme cases". But in our present discussion, what does that mean, "extreme cases"? Of course, we could give this expression many different meanings. I propose to mean "extremely simple". To start with, let's take counting, that is to say, the natural numbers.

These numbers were, of course, a discovery in mathematics. It was a discovery that became very important, indeed essential, in many parts of physics and other sciences. For instance, one counts the clicks of a Geiger counter. One counts the number of white cells under a microscope. Yet the original discovery or invention of counting was not intended for use in science; indeed, there was no "science" at that early date of human culture.

So let us take this possibly childish example, and ask the same question we might ask about a fancier, more modern example. What explains this luck or accident, that a discovery in "pure mathematics" turns out to be good for physics?

Whether we count and find the planets seven, or whether we study the n-body problem, where n is some positive integer, we certainly do need and use counting – the natural numbers – in physics and every other science.

This remark seems trivial. Such is to be expected in the extreme cases. We do not usually think of arithmetic as a special method or theory, like tensors, or groups, or calculus. Arithmetic is the all-pervasive rock bottom essence of mathematics. Of course it is essential in science; it is essential in everything. There is no way to deny the obvious fact that arithmetic was invented without any special regard for science, including physics; and that it turned out (unexpectedly) to be needed by every physicist.

We are therefore led again to our central question, "How could this happen?" How could a mathematical invention turn out, unintentionally, after the fact, to be part of physics? In this instance, however, of the counting numbers, our question seems rather lame. It is not really surprising or unexpected that the natural numbers are essential in physics or in any other science or non-science. Indeed, it seems self-evident that they are essential everywhere. Even though in their development or invention, one could not have foreseen all their important uses.

So to speak, when one can count sheep or cattle or clam shells, one can also count (eventually) clicks of a Geiger counter or white cells under a microscope. Counting is counting. So in our first simple example, there really is no question, "How could this happen?" Its very simplicity makes it seem obvious how "counting in general" would become, automatically and effortlessly, "counting in science".

Now let's take the next step. The next simplest thing after counting is circles. Certainly it will be agreed that the circle is sometimes useful. The Greeks praised it as "the heavenly curve". According to Otto Neugebauer, "Philosophical minds considered the departure from strictly uniform circular motion the most serious objection against the Ptolemaic system and invented extremely complicated combinations of circular motions in order to rescue the axiom of the primeval simplicity of a spherical universe" (*The Exact Sciences in Antiquity*). I. B. Cohen writes, "The natural motion of a body composed of aether is circular, so that the observed circular motion of the heavenly bodies is their natural motion, according to their nature, just as motion upward or downward in a straight line is like natural motion for a terrestrial object" (*The Birth of a New Physics*).

And here is a more detailed account of the circle in Greek astronomy: "Aristotle's system, which was based upon earlier works by Eudoxus of Cnidus and Callippus, consisted of 55 concentric celestial spheres which rotated around the earth's axis running through the center of the universe. In the mathematical system of Callippus, on which Aristotle directly founded his cosmology of concentric spheres, the planet Saturn, for example, was assigned a total of four spheres, to account for its motion 'one for the daily motion, one for the proper motion along the zodiac or ecliptic, and two for its observed retrograde motions along the zodiac' (E. Grant, *Physical Science in the Middle Ages*).

In recent centuries, a few other plane curves have become familiar. But the circle still holds a special place. It is the "simplest", the starting point in the study of more general curves. Circular motion has special interest in dynamics. The usual way to specify a neighborhood of a given point is by a circle with that point as center.

So we see that the knowledge of circles which we inherited from the Greeks (with a few additions) is useful in many activities today, including physics and the sciences. I suppose this is one reason why 10th grade students are required to study Euclidean geometry.

Again, we return to the same question. How can we explain this "miracle"?

Few people today would claim that circles exist in nature. We know that any seemingly circular motion turns out on closer inspection to be only approximately circular.

Not only that. We know that the notion of a circle is not absolute. If we define distance otherwise, we get other curves. To the Euclidean circle we must add non-Euclidean "circles". If the Euclidean circle retains a central position, it does so because we choose – for reasons of simplicity, economy, convenience, tradition – to give it that position.

We see, then, two different ways in which a mathematical notion can enter into science. We can *put* it there, as Ptolemy put circles into the planetary motion. Or we can *find* it there, as we find discreteness in some aspect or other of every natural phenomenon.

Let's take one last example, a step up the ladder from the circle. I mean the conic sections, especially the ellipse. These curves were studied by Apollonius of Perga (262–200 B.C.) as the "sections" (or "cross sections" as we would say) of a right circular cone. If you cut the cone with a cutting plane parallel to an element of the cone, you get a parabola. If you tilt the cutting plane toward the direction of the axis, you get a hyperbola. If you tilt it the other way, against the direction of the axis, you get an ellipse.

This is "pure mathematics", in the sense that it has no contact with science or technology. Today we might find it somewhat impure, since it is based on a visual model, not on a set of axioms.

The interesting thing is that nearly 2,000 years later, Kepler announced that the planetary orbits are ellipses. (There also may be hyperbolic orbits, if you look at the comets.)

Is this a miracle? How did it happen that the very curves Kepler needed to describe the solar system were the ones invented by Apollonius some 1800 or 1900 years earlier?

Again, we have to make the same remarks we did about circles. Ellipses are only approximations to the real orbits. Engineers using earth satellites nowadays need a much more accurate description of the orbit than an ellipse. True, Newton proved that "the orbit" is exactly an ellipse. And today we reprove it in our calculus classes. In order to do that, we assume that the earth is a point mass (or equivalently, a homogeneous sphere). But you know and I know (and Newton knew) that it is not.

Kepler brought in Apollonius's ellipse because it was a good approximation to his astronomical data. Newton brought in Apollonius's ellipse because it was the orbit predicted by his gravitational theory (*assuming* the planets are point masses, *and* that the interactive attraction of the planets is "negligible"). Newton used Kepler's (and Apollonius's) ellipses in order to justify his gravitational theory. But what if Apollonius had never lived? Or what if his eight books had been burned by some fanatic a thousand years before? Would Newton have been able to complete his work?

We can imagine three different scenarios: (1) Kepler and Newton might have been defeated, unable to progress; (2) they might have gone ahead by creating conic sections anew, on their own; (3) they might have found some different way to study the dynamics of the planets, doing it without ellipses.

Scenario three is almost inconceivable. Anyone who has looked at the Newtonian theory will see that the elliptic trajectory is unavoidable. Without Apollonius, one might not know that this curve could be obtained by cutting a cone. But that fact is quite unnecessary for the planetary theory. And surely somebody would have noticed the connection with cones (probably Newton himself).

Scenario one, that Newton would have been stuck if not for Apollonius, is *quite* inconceivable. He, like other mathematical physicists since his time, would have used what was available and created what he needed to create. While Apollonius' forestalling Kepler and Newton is remarkable and impressive, from the viewpoint of Newton's mechanics, it is inessential. In the sequence of events that led to the Newtonian theory, what mattered were the accumulation of observations by Brahe, the analysis of data by Kepler, and the development by Barrow and others of the "infinitesimal calculus". The theory of the conic sections, to the extent that he needed it, could have been created by Newton himself. In other words, scenario two is the only believable one.

If a mathematical notion finds repeated use, in many branches of science, then such repeated use may testify to the universality, the ubiquitousness, of a certain physical property – as discreteness, in our first example. On the other hand, the use of such a mathematics may only be witness to our preference for a certain picture or model of the world, or to a mental tradition which we find comfortable and familiar. And also, perhaps, to the amiability or generosity of nature, which allows us to describe her in the manner we choose, without being "too far" from the truth.

What then of the *real* examples – matrices, groups, tensors, fiber bundles, connections. Maybe they mirror or describe physical reality – "by accident", so to speak, since the physical application could not have been foreseen by the inventors.

On the other hand, maybe they are used as a matter of mere convenience – we understand it because we invented it, and it works "well enough".

Maybe we are not even able to choose between these two alternatives. To do so would require knowledge of how nature "really" is, and all we can ever have are data and measurements, theorems and hypotheses in which we put more or less credence.

And it may also be deceptive to pose the two alternatives – true to nature, like the integers, or an imposed model, like the circle. Any useful theory must be both. Understandable – i.e., part of our known mathematics, either initially or ultimately – and also "reasonably" true to the facts, the data. So both aspects – man-made and also faithful to reality – must be present.

These self-critical remarks do not make any simplification in our problem.

The problem is, to state it for the last time, how is it that mathematical inventions made with no regard for scientific application turn out so often to be useful in science?

We have two alternative explanations, suggested by our two primitive examples, counting and circles. Example one, counting, leads to explanation one: That certain fundamental features of nature are found in many different parts of physics or science; that a mathematical structure which faithfully captures such a fundamental feature of nature will necessarily turn out to be applicable in science.

According to explanation two, (of which the circle was our simple example), there are several, even many different ways to describe or "model" mathematically any particular physical phenomenon. The choice of a mathematical model may be based more on tradition, taste, habit, or convenience, than on any necessity imposed by the physical world. The continuing use of such a model (circles, for example) is not compelled by the prevalance of circles in nature, but only by the preference for circles on the part of human beings, scientists in particular.

What conclusion can we make from all this? I offer one. It seems to me that there is not likely to be any universal explanation of all the surprising fits between mathematics and physics. It seems clear that there are at least two possible explanations; in each instance, we must decide which explanation is most convincing. Such an answer, I am afraid, will not satisfy our insistent hankering for a single simple explanation. Perhaps we will have to do without one.

Bibliography

1. Bernard Cohen, *The Birth of a New Physics* (Doubeday & Company, 1960).
2. Edward Grant, *Physical Science in the Middle Ages* (Cambridge University Press, 1977).
3. Michael Monastyrsky, *Riemann, Topology, and Physics* (Birkhäuser Boston, 1987).
4. O. Neugebauer, *The Exact Sciences in Antiquity* (Dover Publications, Inc., 1969).
5. Mark Steiner, *The Application of Mathematics to Natural Science, J. Philosophy*, **86** (1989) pp. 449–480.
6. Eugene Paul Wigner, *Symmetries and Reflections* (Bloomington Indiana University Press, 1967).

Mathematics and the Natural Order

Wendell G. Holladay
Department of Physics and Astronomy
Vanderbilt University
Nashville, TN 37235, USA

1. The Nature of Mathematics

Since Plato, it has not been uncommon to regard mathematics as composed of divine, eternal, perfect, absolute, certain, infallible, immutable, necessary, *a priori*, exact, and self-evident truths of ideal forms existing in their own world.[1–4] A major question with this Platonist view of mathematics is how this world of ideal forms relates to the everyday world of human experience, where for several millennia mathematics has held a pre-eminent position.

An alternative to this Platonist view of mathematics is the formalist position, promoted especially in the early part of this century by the great German mathematician David Hilbert.[5] In its vulgar form, to use Browder's phrase, formalism "was taken to say that mathematics consists simply of the formal manipulation of uninterpreted symbols".[6] This formalist doctrine, almost pristine in its detached purity, too is difficult to reconcile with the great significance and power of mathematics in human affairs[7] and with its evolutionary roots in human experience with nature.[8] For example, the idea of natural numbers (the integers) undoubtedly arose as a result of counting objects in the natural world – two eyes, ten fingers, six goats, etc. Wilder[9] points out that some form of counting is found in all primitive cultures.

It is worth noting that counting would be difficult in a world where, without exception, entities are nonseparably intertwined or entangled with one another, i.e., where there is no distinct individuation of either objects or events. Remarkably, as will be seen later, our world appears to be possessed of both nonseparable characteristics as well as distinct features of individuality and identity. Locating and understanding the roots of this peculiar dualism are a challenge.

In our world, events as well as objects can be counted – eight steps across the river, four sunrises, six "moons", etc. Experience with these numbers naturally leads to the processes of arithmetic – addition, multiplication, subtraction and division, from which such concepts as fractional and negative numbers arise. The arithmetic of events and the important introduction of a unit of length – a step, the length of the forearm, notches on a stick, or knots on a rope – allow geometry (land measurement) to be developed and temples and tombs (pyramids) to be constructed.

The extraordinary properties of visual perception of normal human beings must play a crucial role in the origins of geometry. The visual field of humans (and presumably other animals with binocular vision as well) reproduces with a considerable degree of fidelity the actual spatial relations that exist in nature. Our perception of the near and the far, the up and down, the left and right and a host of shapes, forms, and relations reflect with remarkable accuracy spatial relationships that can be confirmed by much other evidence, both qualitative and quantitative.

The further development and refinement of geometry by the ancient Greeks culminated in the abstract axiomatic system presented in Euclid's *Elements* in about 300 B.C., surely the most widely read and influential textbook ever written. Our genetic origins seem to be African, our religious heritage largely Oriental, but modern mathematics, science, and technology originated in the West, for which Euclid's geometry undoubtedly bears a major responsibility.

Euclid's great treatise on geometry had, of course, included the Pythagorean theorem and the discovery that the diagonal of a square is incommensurate with the length of a side. To state this result another way, the diagonal length of a unit square is $\sqrt{2}$, an irrational number. Eudoxus' theory of proportions "rationalized" the existence of irrational numbers, and thus laid the foundations for a rigorous treatment of the real number continuum, finally achieved only in the 19th century by Dedekind and Weierstrass.

The sciences of number and space provide the seminal models that have guided and inspired the subsequent development of a network of mathematics, including analytic geometry, calculus, complex analysis, non-Euclidean geometries, topology, modern algebra, number theory, set theory, logic and statistics.[10] These developments have involved the processes of extraction, consolidation, generalization, distillation, symbolization and formalization[11] of a content that transcends "uninterpreted symbols". This content has been characterized by Hardy,[12] Whitehead[13] and Steen[14] as a study of patterns, by Schneer[15] as a study of order and relationship, by Courant and Robbins[16] as a study of structure and relationship, by MacLane[17] as a study of significant forms, and by Putnam[18] as a study of possible abstract structures. According to Browder,[19] Descartes and Leibnitz held views about mathematics similar to these.

This characterization of mathematics as a study of the order and relation of possible abstract patterns, forms, and structures has several advantages. It dispenses with the heavy baggage of a separately existing Platonic world of ideal forms and makes mathematical knowledge possible to humans without supernatural revelation. It allows for the dialectical development of mathematics stimulated by the cultural environment and especially by the order and structure extracted from nature; yet it frees mathematics from empirical bondage and provides mathematics with an independent content, formal structure and methodology of its own. Thus it provides mathematics with root and substance, yet allows for the full range of the intuitive,[20] creative and aesthetic impulse for which mathematics is so justly renowned. It is consistent with Halmos' analysis of mathematics as a creative art.[21]

2. Mathematics and Nature

Arithmetic and geometry – concerned respectively with number and space – each presumably originated from direct experience with form and structure experienced in nature. It is not surprising therefore that mathematics is a key element in grasping the order and structure of nature. What is noteworthy is the extent – the scope and depth – to which mathematics, generalized and formalized, succeeds in doing this. A list of some major historical figures who were seized with this vision will indicate its persistence and power[22–26]: Pythagoras (6th c B.C. Greece), Plato and Eudoxus (4th c B.C. Greece). Archimedes (3rd c B.C. Greece), Grosseteste (13th c English), Nicholas of Cusa (15th c German), Copernicus (15th c Polish-German), Kepler (17th c German), Galileo (17th c Italian), Descartes (17th c French), Newton (17th c English), Euler (18th c Swiss), D'Alembert (18th c French), Lagrange (18th c French-Italian), Gauss (19th c German), Riemann (19th c German), Cauchy (19th c French), Hamilton (19th c Irish), Maxwell (19th c Scot), Poincaré(19th c French), Hilbert (20th c German), Einstein (20th c German), Eddington (20th c English), Weyl (20th c German-American), Dirac (20th c English), Schrödinger (20th c Austrian), von Neumann (20th c Hungarian-American), Wigner (20th c Hungarian-American), Yang (20th c Chinese-American).

Spread as they are across a time period of over 2.5 millennia, it is not surprising that these thinkers do not always speak with one voice, but their common commitment to the idea of the existence of an intimate bond between mathematics and nature is clear. One major theme in this bond is the use of mathematics as a tool in describing nature. The practice of measurement, discussed earlier, exemplifies the use of numbers and geometrical patterns as tools for dealing with some of our most basic experiences with nature. A few other examples are worth considering.

Experience with repetitive events – the rising and setting sun, the new moon, and the seasons – allowed time to be conceived, tracked, measured, and recorded. Long before Euclid, the Mesopotamians had developed a sophisticated sexagesimal (base 60) number system for recording the angular positions of celestial bodies and for calculating the occurrence of the cyclical events mentioned above, and others as well, especially lunar and solar eclipses. We retain elements of the sexagesimal number system in the units of second and minute for tracking time as well as for divisions of an angular degree. It is not accidental that each angle of an equilateral triangle has 60°.

These sterling accomplishments of Babylonian astronomy and mathematics became available to the Hellenistic world by Alexander's conquest in about 335 B.C.. In that new intellectual environment, the influx of new ideas flourished and stimulated the development of the science of trigonometry for the management of angular relationships. Greek astronomy was also richly fertilized and produced such impressive results as:

1. Hipparchus' (~125 B.C.) assessment of the length of the solar year (the time interval between, say, two successive winter solstices, when the sun is lowest in the sky) as 365 days, 5 hours, 55 minutes, and 12 seconds, too long as compared to current measurements by only 6 minutes and 26 seconds.
2. Hipparchus' discovery of the precession of the equinoxes with a period of about 26,000 years.
3. Ptolemy's comprehensive geocentric system (about 150 A.D.) for reproducing the position of celestial objects in the sky as seen from the Earth.

Another example of the use of mathematics in describing nature comes from the immensely important work of Archimedes (~200 B.C.). He showed that a circle of radius r in a 2-dimensional Euclidean plane encloses an area of πr^2, (where π is the ratio of the circumference to the diameter of the circle) and the volume enclosed by a sphere is $(4/3)\pi r^3$. He found an accurate method to calculate the value of π, analyzed simple machines (e.g. the lever) and captured the principle of hydrostatic buoyancy. The methods used by Archimedes to achieve these results have remained essential to the development of both mathematics and physics.

An adequate description of motion eluded the ancients. The modern concepts of velocity as the time rate of change of position and of acceleration as time rate of change of velocity are needed for a proper mathematical description of motion. These ideas were beginning to germinate in the minds of 14th c scholars in Oxford (Bradwardine, Heytesbury, and Swineshead) and Paris (Buridan and especially Nicholas Oresme).[27]

The further development of these concepts by the use of convenient algebraic symbolism derived from Hindu and Arabic sources and especially through measurements of bodies moving in controlled experiments allowed Galileo (~1600) to express succinctly and powerfully his law of freely falling bodies, $d = (1/2)gt^2$, where d is the distance of fall from rest, t is the time of fall and g is the acceleration due to gravity. At about the same time Kepler found his famous third law of planetary motion that the period T of a planet's revolution about the sun and its distance a from the sun followed the relation $T^2 = Ka^3$, where K is a constant for all the planets (which Newton showed, after Kepler, depends on the mass of the sun). Kepler had earlier enunciated his first two laws of planetary motion that established the shape of the orbits as ellipses, a geometrical figure whose properties had been worked out 1800 years earlier by Apollonius in his study of conic sections (the circle, ellipse, parabola and hyperbola).

By the 17th c, a sufficient conceptual foundation existed on which Newton could build a comprehensive mathematical theory of motion. With a specific set of ideas about space, time, mass, and force (specifically the universal law of gravitational force between two masses) and the idea of mathematical function,[28] Newton's system embraced all that was known (and more) about terrestrial and celestial motion. The calculus, which he and Leibnitz independently conceived, became the natural and versatile language for structuring and expressing the mathematical principles of motion. Newton's system provided the means for understanding and demystifying comets and for calculating in the mid-19th c the position of an unknown planet that seemed to be affecting the motion of Uranus. A telescope aimed at that point found the planet, Neptune. The power of Newton's system for analyzing motion was greatly expanded in structure, generality, and applicability by the analytic virtuosity of such masters as Euler, Lagrange, Laplace, Gauss, Cauchy, Hamilton, and Poincaré.

These explicit examples, taken from ancient and more recent times, serve to illustrate how mathematics has been used to describe, analyze, systematize, interpret, investigate, discover, and formulate the laws of nature. They reveal a link between mathematics and the natural order and serve as a partial fulfilment of Descartes' vision of mathematics as the key to unlock nature's secrets.[29] They exemplify Wigner's penetrating insight of the "miracle of the appropriateness of the language of mathematics for the formulation of the laws of physics".[30] Whether as handmaiden or Queen of the sciences, mathematics, says Bochner, "is an indispensable medium by which and within which science expresses, formulates, continues and communicates itself."[31] Whitehead opts for Queen: "The utmost abstractions (of mathematics) are the true weapons with which to control our thought of concrete fact."[32]

However, there runs through the ages another major theme of deeper significance that transcends these largely utilitarian purposes of mathematics. This theme that mathematical structure and harmony is the essence of order and relationship in nature is captured in the 6th B.C.. Pythagorean metaphor that the world is made of number (interpreted by Burtt to mean geometrical units) and has remained a powerful strand in human thought from Plato through the 15th c German Cardinal Nicholas of Cusa, Copernicus and his teacher de Novara, Kepler, Galileo, Leibnitz and on to the present century in Jeans, Eddington, Weyl, Dirac and Einstein. For many of these authors, commitment to mathematical harmony in nature is closely linked to transcendent design. For example, hear Plato: God ever geometrizes.[33] And the great Cardinal Nicholas, no doubt inspired by Almighty power, held that "the world is an infinite harmony in which all things have their mathematical proportions".[34] Kepler (God made the world according to number),[35] Leibnitz (As God calculates, so the world is made),[36] Jeans (The Great Architect of the Universe now begins to appear as a pure mathematician)[37] and Weyl agree on the divine role in the mathematical design of the Universe.[38] A similar conviction was likely behind Galileo's celebrated statement: The book (of the Universe) is written in mathematical language.[39]

The mathematical harmony of nature can, of course, be discussed with theological neutrality, as Einstein does in his statement: "Our experience hitherto justifies us in believing that nature is the realization of the simplest conceivable mathematical ideas."[40] One could question the aptness of the word "simplest", even in Einstein's own work, but his commitment to mathematical order and structure in nature is evident.

What are the natural grounds that warrant the commitment to the notion of a nature possessed of mathematical structure? First, the notion provides the rationale for understanding the origins of mathematics itself. According to Sullivan,[41] the whole of mathematics is built up from two concepts, number and space. The roots of these concepts, says Courant, "reach down... deep into what might be called reality."[42] These positions are wholly consistent with the nature of mathematics as presented in Sec. 1, i.e., as a study of order, relations and patterns of possible abstract structures but now rooted in and inspired by nature itself.

Second, the mathematical structure of nature explains the great utility of mathematics in describing, analyzing and mastering nature, for mathematics, or at least much of it, is distilled, abstracted, and generalized from nature itself. Thus, the role of mathematics as Descartes' "key to unlock nature's secrets" or as Wigner's "miracle" for formulating the laws of physics, or as Bochner's "medium in which science expresses itself", or as Whitehead's "true weapon with which to control our thought of concrete fact" becomes demystified and rationally explicable, without any loss of autonomy or flexibility of mathematics itself.

Finally, the strongest support for belief in the mathematical harmony and essence of nature is based on the stunning consequences that come from the mathematical theories of nature. Maxwell's equations of electromagnetism provide a unified account of electromagnetism and light. In Maxwell's day it was generally supposed that his equations described disturbances in the ether, but Hertz took the seminal position that these equations were about electromagnetic fields in space and time and that electromagnetic theory itself was these equations.[43] Hertz produced and detected the radio waves that were directly implied by the mathematical structure of the equations, a discovery that made possible today's vast communications industry of radio and television. Microwaves, radar, infrared, ultraviolet, X-rays, and γ-rays are all electromagnetic fields embraced within the structure of Maxwell's equations. Moreover, the space-time relationships between moving systems were deduced by Lorentz from these equations, which, Einstein realized, implied that the speed of light (electromagnetic waves) c is a constant (about 300,000,000 meters per second) independent of relative motion. Einstein further deduced the wholly unexpected result that the time duration between ticks of a moving clock would expand ("time on a moving clock slows down") and that energy and inertial mass are equivalent ($E = mc^2$), both results since experimentally validated to a high degree of precision. Maxwell's equations seem to incorporate and reveal fundamental features of mathematical structure in nature.

Einstein's equations of general relativity have a similar standing. By combining the space-time structure implied by Maxwell's equations with the equivalence of an acceleration and a gravitational field (accelerating under free fall in a gravitational field cancels the effect of the gravitational field – i.e., an object becomes "weightless"), Einstein developed an elegant 4-dimensional, non-Euclidean (Riemannian) space-time geometrical theory of gravitation. Quantitative predictions of the bending of a light ray in a gravitational field (star light passing the eclipsed sun) and the motion of an object in a strong gravitational field (the orbit of the planet Mercury about the sun) were confirmed. Moreover, Einstein's equations are not consistent with a Universe that is static in space and time, i.e., the equations require that the Universe either expand or contract.[44] In addition, they require that an expanding Universe have a beginning, i.e., a singular point in space-time, the Big Bang. Astronomers have indeed observed that all galaxies appear to be moving apart, those travelling the fastest naturally being the greatest distance from one another. Moreover, radiation permeates the Universe uniformly, precisely consistent with the expected explosive debris from the Big Bang.[44]

Other mathematical structures have unified electromagnetism with another force of nature, the so-called weak force that is manifested in certain radioactive nuclear processes. Moreover, similar, though somewhat more complicated (but impressively elegant), mathematical forms show great promise for representing the strong force that holds quarks in neutrons and protons and binds the latter together to form atomic nuclei.[45]

These mathematical theories of nature are impressive in their aesthetic appeal[46] and show great precision and versatility in representing the very structure of nature itself. But they are neither perfect nor complete. Nature is subtle and does not easily reveal its secrets, a characteristic that will become more apparent when another great mathematical structure, quantum theory, is discussed below in Sec. 4.

3. Scientific Realism

The position that mathematical order and relationship exist in nature and are approximately embraced by the mathematical theories of physics is a particular aspect of a world view called realism. The most basic tenet of realism holds that a world exists independent of the human mind. This general statement leaves open the possibility of many refinements or variations, including scrutiny of the question – What all is in this world? Objects? Fields? Systems? Relations? Processes? Universals? Minds? Spirits? Actualities? Possibilities? etc.

There are many "-isms"[47] – Kant's critical idealism, phenomenalism, pragmatism, instrumentalism, descriptivism, operationalism, for example

– that accept the existence of a mind-independent world but deny that humans can know much about it in itself. Humans may know their individual experiences, but the extent to which these experiences, and reflections on them, reveal anything about the world is, according to these "-isms", problematic, the degree of skepticism varying among the "-isms".

However, epistemological realism in general not only adopts the view that a world independent of the human mind exists but that humans can achieve knowledge of it. A *naive* realist holds this position and points to the immediate objects of experience as evidence for it. Scientific realists[48,49] are more sophisticated, critical, probing and specific. They claim that terms in scientific theories refer to entities or processes in the mind-independent world of nature (i.e., in reality), and that the knowledge obtained by the processes and canons of science is approximately true of this reality.

What are the grounds for these claims of scientific realism? One important argument, due to Putnam,[50] states that, if scientific theories were not approximately true, then their success would be a miracle. Wigner does speak of a "miracle" in the relation of mathematics to physics, and, it is not irrational to hold that the *existence* of something (anything) compared to its *nonexistence* can be deemed a miracle. But according to Boyd,[51] the deficiency of the miracle argument is that, while it may support adherence to the doctrine of realism, it does not diagnose what is wrong with other competing positions such as Kuhn's argument that theories are constructed with incommensurable paradigms and the standard empiricist position that empirically equivalent theories are evidentially indistinguishable.

Boyd, however, seeks to remedy this deficiency by a detailed exposition to the effect that scientific realism "provides the only scientifically plausible explanation for the instrumental reliability of the scientific method".[52] First, the claim that scientific theories embrace approximate truths about nature does provide rational grounds for instrumental reliability. Under these circumstances it becomes understandable how scientific theories can predict previously undetected behavior and structures in nature such as the planet Neptune, radio waves, the equivalence of energy and mass, etc. Moreover, the application of the theory to guide the development of new and sophisticated technologies that perform according to their scientifically designed specifications simply becomes standard procedure.

Boyd then argues that neither the constructivist nor the empiricist can explain the instrumental reliability of science. Constructivism fails because instrumental reliability "cannot be an artifact of the social construction of reality...it cannot be that the explanation of the fact that airplanes, whose design rests upon enormously sophisticated [scientific] theory, do not crash is that the paradigm *defines* the concept of an airplane in terms of crash resistance".[53] A similar statement could be made about electric generators, radios, computers, and nuclear reactors. Empiricism fails according to Boyd because the evidential indistinguishable thesis is false. It is false because in assessing the plausibility of a proposed theory it ignores theory-mediated evidence which "is no less empirical than more direct experimental evidence – largely because the evidential standard that apply to the so-called direct experimental tests of theories are theory-determined in just the same way that judgments of plausibility are".[54] In other words, theory and experiment are so intertwined that the concept of "empirically equivalent theories" is ambiguous.

These arguments for scientific realism are important with respect to the discussion earlier in this paper about mathematical order and relationship in nature, for if scientific realism (the doctrine that the theories of science are approximately true) were undermined, then the claim that the mathematical structures of the great theories of physics reflect approximately the mathematical order and form of nature itself would be similarly undermined. It would then follow that the idea of mathematics as patterns and possible structures grounded in, or generalized from, the form and structure experienced in nature would lose its rational foundations. A different origin of mathematics, or at least a different rationale for its origin, would have to be found and the "unreasonable effectiveness" of mathematics in physics would become a major conundrum. On the other hand, the validity of scientific realism would sustain the historic claim that mathematical structures of theoretical physics embrace (approximately) the forms and relations of nature itself, for these mathematical structures are essential elements of the theories. The roots of mathematics as inspired by nature and its effectiveness in science would fall into place.

4. Mathematical Structure in the Quantum World

The mathematical structure of nature is apparently manifest in such theories of physics as electromagnetic theory and general relativity. These theories are characteristic of classical physics in that the physical quantities always have definite values and follow deterministic laws. To be discussed in this section is the nature of the mathematical structure of quantum theory. For a system with speed small compared to the speed of light, these mathematical structures are the wave states that are mathematical solutions of the nonrelativistic Schrödinger equation.

The issue to be addressed is whether, or to what extent, mathematical structures can be taken as structures of nature as compared to the notion that they are merely mathematical expressions for correlating one set of experimental results with another set. A case can be argued for both positions. Consider first the claim that the mathematically expressed wave states of quantum theory reflect the existence of such structures in nature. Atomic and molecular structure provides especially strong examples.[55]

First, the permanence or stability of the atoms and molecules (as well as nuclei, protons and other structures) are fully contained within the mathematics of quantum theory. The wave structure corresponding to the permanent, lowest energy solution is well-defined, and for simple systems such as the hydrogen atom, can be very accurately calculated and found to be in full accord with experimental evidence.

Second, the periodic properties of the chemical atoms as exhibited in the periodic table of the elements can be reproduced by the mathematical solutions of the Schrödinger equation, provided there is added the requirement of the Pauli exclusion principle that no two electrons can occupy the same wave state. This requirement in turn follows as an essential element of the mathematical structure of relativistic quantum field theory.

Third, the detailed structure of individual atoms and molecules reflect the mathematical solutions to Schrödinger's equation. For example, according to this equation the stable wave state of methane (CH_4) is a tetrahedral structure composed of a hydrogen atom at each of the four corners, with the carbon atom in the center.[56] This is the measured structure of stable methane. The structure of other molecules such as water (H_2O), ammonia (NH_3), etc. follows directly from the wave state solutions of the Schrödinger equation for these systems.

Fourth, according to the Schrödinger equation the four outer electrons in the carbon atom form a tetrahedral structure that should be present in the solidified crystalline form of this substance. Such is indeed the measured structure of diamond.[57] In fact, the angles between the crystalline planes of many crystals can be perceived directly by the naked eye and these features reflect the structure of the atoms themselves as contained in the mathematical solutions of Schrödinger's equation. Indeed, no exception is known to the claim that all atomic nuclei, nucleons freed from nuclei, atoms, molecules, and perfect crystals reflect in precise detail the mathematical structure of quantum theory.

An example of a different kind (and an unusually impressive one as well) that exhibits the underlying structure of nature is the Dirac equation for the electron, which automatically contains within its mathematical structure the correct value of the spin of the electron. Even more remarkably, the Dirac equation for the electron embraces the anti-electron (positron). That the Dirac equation has this property was recognized before the positron was experimentally detected. Moreover, according to this equation, antiparticles should exist for any particle that has the same spin as the electron, and many such particle-antiparticle pairs (proton-antiproton, neutron-antineutron, etc.) have been detected. This capacity to include anti-matter in the mathematical theories originally constructed only for ordinary matter must be regarded as one of the truly stunning triumphs of mathematical physics.

A final example of the astonishing success exhibited by the mathematical structure of quantum theory is that feature of the theory that represents the holistic structure of physical systems, i.e., that property of a physical system wherein it behaves as an organic whole without localized, individuated segments. The wave states of quantum theory possess this property of wholeness both for one particle states, say of an electron, a photon, or a proton, as well as for multiparticle states characteristic of nuclei, atoms, molecules, crystals, and photons. This holistic property of one particle states is revealed by interference patterns[58] that become apparent in circumstances where one part of a one particle wave state may add to or subtract from another part of the wave state, just as water waves do in a ripple tank.

The indivisible wholeness of the wave state has the concomitant characteristic of indefiniteness[59] in the values of certain physical quantities such as the position and momentum of the particle, though the mass and electric charge of the particle each has a definite value. In more technical language the indefinite nature of the quantum wave state of a physical system can be represented by a linear superposition of states, each of which has a precisely defined value of some physical quantity. Under appropriate conditions, a component of these superposed states may combine with other components to produce interference patterns that reflect characteristics of the constituent states.

The holistic nature of multiparticle systems is revealed in various ways. An ordinary oxygen molecule (O_2) is composed of two oxygen atoms, each with 8 protons and 8 neutrons in its nucleus, the two nuclei being surrounded by a total of 16 electrons. The stable wave state of this fairly complex system of 48 entities behaves as a unit, without localized, individual parts. The behavior of the whole in all of its aspects is not equal to the sum of the behavior of the individual parts. The stable wave state of one such O_2 molecule is identical to that of any other O_2 molecule. Similar statements can be made about other molecules, atoms, nuclei and perfect crystals.

Moreover, the holistic indivisibility of wave states can extend beyond atomic dimensions and include vast regions of space. For stellar photons the Hanbury Brown-Twiss effect manifests such long distance correlations.[60] Moreover, Aspect *et al.*, have measured correlations at laboratory distances between two photons that agree with the nonseparable and indefinite structure of quantum mathematics and *violate* relations (Bell inequalities) that are derived on the assumption that physical quantities associated with the two photons have spatially local and definite values.[61–65] Thus the value of these quantities is neither local nor definite. The mathematical structure of the wave states of quantum theory incorporates the truly remarkable indefinite, indivisible, nonseparable features of nature.

Presumbly, a world governed only by these quantum characteristics – i.e., possessed only of mathematical structures characterized by indefinite, indivisible, and nonseparable wholeness — akin perhaps to the permanent One of Parminides — would be bereft of localized individualistic objects and of the occurrence of specific, recorded events. An intelligence in such a world could hardly find anything in it that corresponds to everyday experience with separate identifiable entities and events. In such a world the elementary processes of counting based on collections of individuals and even the recognition of relations in space and time appear problematic.

As sure as is our knowledge of the mathematical structure of the quantum world, our knowledge of the world of everyday experience, the classical world of distinct objects and events in nature, is even more certain. But these worlds appear to be incompatible. A specific example, due to Einstein,[66] highlights clearly this paradoxical dichotomy of nonseparable indefiniteness on the one hand, and the explicit distinctiveness of things on the other. A system, say an excited atom or a radioactive nucleus, is confined in a certain region of space. The excited atom then emits a photon or the radioactive nucleus emits a material particle, say an electron. The wave state of the emitted object propagates in space according to the standard mathematical equation of quantum theory, the Schrödinger time dependent equation. The wave pattern can be discerned by an interference type experiment that reveals the distribution in space-time of the wave state.

Yet, when the "wave-photon" is actually detected, it will deposit all of its energy in a single act at a localized region of space, perhaps on a single silver bromide grain in a photographic plate or on a single electron in a photoelectric detector (this is how photoelectric detectors actually work). Similarly, the electron from a radioactive nucleus will propagate in space-time with an extended, well-defined mathematical structure, but will produce a distinct path somewhere on a photographic plate or in some other device such as a bubble chamber that can record the track. The probability of detecting the proton or the electron in any specific, localized region of space depends on the absolute square of the amplitude of the propagating wave in that particular localized region. When the localized recording of the photon absorption or the electron track occurs, the propagating wave disappears everywhere else. The wave-photon or wave-electron propagates smoothly and continuously; on the other hand, the energy carried in it is deposited in one single, localized, definitive event. Contrast this behavior with a water wave created by dropping a 50 pound rock from a height of 20 feet onto the surface of a smooth lake. The wave propagates outward from the point of impact and its energy is eventually absorbed all along the lake shore by innumerable small processes. If the water wave behaved as the "photon-wave" does, the wave would spread out as before, but the total energy would be absorbed in one single indivisible act by knocking just *one* of many 50 pound rocks lining the shore 20 feet into the air.

The atom that emits the wave-photon of light could be on the surface of the sun or some other star and the wave could spread over vast regions of space, yet the total energy of the wave-photon could be absorbed in a spatial region of atomic dimension, say by the chlorophyll in the leaf of a plant in my yard. Solar photons, in fact, carry the energy for the growth of plants and the production of food on earth. The fossil record reveals that this process has a history spanning billions of years. The point is that the localized absorption of the spreading wave of a solar or stellar photon is a natural process that has been occurring for eons, requiring no human instruments or human intervention of any sort.

The extraordinary situation seems to exist that elegant mathematical forms exhibit holistic structural features in quantum states of matter and light. These wave-states are time reversible, i.e., replacing t (the time variable) by $-t$ preserves the structure of these equations. But actual events with separable, identifiable single entities do occur, with the recording system connected with the wave states only in a probabilistic and time irreversible sense. How is this chasm between two such disparate modes to be bridged? How does an indefinite, indivisible quantum world ever exhibit things with individual, definite identity? How in such a world does anything definitive ever happen at all?

Bohr's response to this enigmatic situation has become the established orthodoxy. According to Bohr,[67] a necessary condition for a specific event to occur, for a definitive result to be recorded in a measurement process, requires an arrangement describable in purely classical terms that precludes a separate account of the interaction between the quantum system and the recording arrangement and necessitates a statistical mode of description for a particular, individual outcome. The full meaning of the quantum wave state is revealed only by various complementary, mutually incompatible, arrangements of classical systems. For Bohr, the mathematical formalism of quantum theory provides symbolic tools for calculating probabilities that this or that quantum event will be recorded under well-defined arrangements "specified by classical physics concepts".[68]

Bohr's perceptive analysis has the great advantage of providing a faithful, reproducible and extraordinarily successful description of an enormous range of experimental phenomena. However, his position poses serious problems, for it:

1. Implies, according to von Weizsäcker,[69] "an apparent paradox: classical physics has been superseded by quantum theory; quantum theory is verified by experiments; experiments must be described in terms of classical physics";
2. Yields no insight into how the holistic, nonseparable, indefinite quantum world suddenly manifests separate, distinct, individual entities with definite values of certain physical quantities when a classical system is appropriately arranged. The questions of how a classical system ever becomes established in the first place and how definitive events ever actually occur remain unanswered, except by fiat;
3. Leaves unanswered, except by fiat, the question of under what circumstances and for what physical systems the time reversible Schrödinger equation is valid and why it is not valid for the physical systems that record time irreversible events, such as occur in measurement and

other event-sensitive processes;[70] and

4. Treats the mathematics of quantum systems as a symbolic tool for calculating probabilities that this or that result actually occurs. While this approach has its undeniable success, it fails to elucidate either any underlying rationale of why the mathematical tools work so well or any deeper understanding of the mathematical forms of matter and energy specified by the quantum equations and apparently ensconced in nature. According to the scientific realists, the only elucidation available that explains the success of physics (and also explains the roots of mathematics) is that the mathematical structures approximate in some sense the structures of nature. This position is consistent with the great historic tradition going back to Pythagoras of the intimate relationship between mathematics and the natural order.

Bohr and his disciples seem to feel that further analysis will be forever impotent to penetrate more deeply into these issues. Nevertheless, many have sought deeper understanding of the singularly peculiar dual circumstance of a classical world in which physical quantities are definite and events actually occur, probabilistically conditioned by the precise mathematical wave-states of the nonseparable, indivisible quantum world. Von Neumann[71] and Wigner[72] argue for an essential role of the human consciousness to account for this dualism, but a dandelion grows in the quad by localized absorption of a space extended wave-photon without any help at all from a consciousness (unless, of course, God is in the quad). Others have sought a mathematical structure that would treat with coherent unity both the quantum and the classical worlds. Bohm[73] has focused on the "quantum potential" to achieve this purpose; Vigier *et al.*,[74] on a "subquantum Dirac ether"; Primas[75] on "algebraic quantum mechanics"; and Penrose[76] on general-relativistic gravitation. Everett[77] and his followers preserve the full mathematical integrity of the orthodox quantum structure but do so at the exorbitant cost of creating a new, disjoint universe (many worlds) for each possibility of a recorded result contained in the original wave state. This ontological profligacy of continually creating untold numbers of new universes at the occurrence of each and every event makes the existence of the Platonic world of ideal forms seem downright parsimonious.

Each of these imaginative proposals brings novel, though different, features into the theory to serve the goal of mathematical unity and coherence. None has yet made such pivotal contact with nature as to compel wide acceptance. The dualism of a quantum and classical world remains unresolved.

5. Summary

The mathematical structures of theoretical physics – electromagnetic theory, relativity, quantum theory – have shown great power, elegance, and versatility in representing the structures of nature itself. The mathematical structures of the quantum world are extraordinarily subtle and fruitful in reflecting the nonseparable and indivisible wholeness that seems to characterize essential elements of the world; yet at the very point of the action – occurrence of actual events in nature – this great mathematical edifice apparently falters and, according to some, becomes an instrumental calculus for probabilistic predictions only. This latter assessment raises a serious challenge to those who seek a more coherent grasp of nature in terms of consistent, mathematical structures. Efforts to meet this challenge within the canons of the long tradition of a faithful mathematical representation of nature have not been very revealing.

But while mathematics and physics with their origins in antiquity have an enviable maturity, many of their achievements are recent and ongoing. They have by no means exhausted their creative potential. New developments and insights can reveal clearer vistas of nature's landscape, and, when they do, past experience suggests that mathematical concepts and structures of even greater generality and abstraction will capture the contours of that landscape.

References

1. L. O. Katsoff, *A Philosophy of Mathematics* (Iowa State, 1948).
2. Stephen Körner, *The Philosophy of Mathematics* (Dover, 1986).
3. Philip Kitcher, *The Nature of Mathematical Knowledge* (Oxford, 1984).
4. In *Philosophy of Mathematics: Selected Readings*, eds. Paul Benacerraf and Hilary Putnam (Cambridge University Press, 1983).
5. John von Neumann, "*The Formalist Foundations of Mathematics*" *Ibid.* pp. 61–65.
6. Felix E. Browder, "*Does Pure Mathematics Have a Relation to the Sciences*", *Am. Sci.* **64** (1976) 542–549.
7. Morris Kline, *Mathematics in Western Culture* (Oxford University Press, 1953).
8. R. L. Wilder, *Evolution of Mathematical Concepts* (Open University, 1968). See also several articles in *Mathematics: People, Problems, Results*, eds. D. M. Campbell and J. C. Higgins (Wadsworth, 1984), vol. I, Part three, *The Development of Mathematics*, pp. 327–404.

9. Wilder, *Op. Cit.*, p. 32.

10. Carl Boyer, *A History of Mathematics* (Princeton University, 1968).

11. R. L. Wilder, *Mathematics as a Cultural System* (Pergamon, 1981).

12. G. H. Hardy, *A Mathematician's Apology* in *The World of Mathematics* (James R. Newman), vol. 4 (Simon & Schuster, 1956), pp. 2027–2038.

13. A. N. Whitehead, *Mathematics and the Good*, in *The Philosophy of Alfred North Whitehead*, ed. P. A. Schilpp (Tudor, 1941).

14. L. A. Steen, *The Science of Patterns*, *Science* **240** (1988) 611–616.

15. C. J. Schneer, *The Evolution of Physical Science* (Grove, 1960), Chap. 6, *The Idea of Mathematics*, pp. 91–102.

16. R. Courant and H. Robbins, *What is Mathematics?* (Oxford University, 1941).

17. Saunders MacLane, *Mathematics: Form and Function* (Springer-Verlag, 1986).

18. Hilary Putnam, *Mathematics, Matter and Method: Philosophical Papers* (Cambridge University, 1975), vol. I, Chap. 4, "*What is Mathematical Truth?*", pp. 60–78.

19. Felix E. Browder, *Mathematics and the Sciences*, pp. 278–292 in *History and Philosophy of Modern Mathematics*, eds. William Aspray and Philip Kitcher (University of Minnesota, 1988).

20. Raymond L. Wilder, *The Role of Intuition*, *Science* **156** (1967) 605–610. Reprinted in *Mathematics: People, Problems, Results*, eds. Douglas M. Campbell and John C. Higgins (Wadsworth, 1984), vol. II.

21. Paul R. Halmos, *Mathematics as a Creative Art*, *Am. Sci.* **56** (1968) 378, *Op. Cit.*, pp. 19–29.

22. Dirk J. Struik, *A Concise History of Mathematics* (Dover, 1967), 3rd edition.

23. Morris Kline, *Mathematics and the Search for Knowledge* (Oxford University, 1985).

24. E. J. Dijksterhuis, translated by C. Dikshoorn, *The Mechanization of the World Picture: Pythagoras to Newton* (Princeton University, 1986).

25. E. A. Burtt, *The Metaphysical Foundations of Modern Science* (Doubleday Anchor, 1954).

26. Salomon Bochner, *The Role of Mathematics in the Rise of Science* (Princeton University, 1966).

27. Dijksterhuis, *Op. Cit.*, pp. 179–200; Boyer, *Op. Cit.*, pp. 288–292.

28. Spengler called the idea of function "The Symbol of the West". Oswald Spengler, *The Meaning of Numbers* in James R. Newman, *The World of Mathematics* (Simon & Schuster, 1956), vol. 4, p. 2334.

29. E. A. Burtt, *Op. Cit.*, p. 204.

30. E. P. Wigner, *The Unreasonable Effectiveness of Mathematics in the Natural Sciences* in *Symmetries and Reflections* (Indiana University, 1967), p. 237.

31. S. Bochner, *Op. Cit.*, p. 256.

32. A. N. Whitehead, *Science and the Modern World* (The Free Press, 1925), Chap. 2, *Mathematics as an Element in the History of Thought* reprinted in Newman, *The World of Mathematics* (Simon & Schuster, 1956), vol. 1. The reference to mathematics as Queen of the sciences is attributed to Gauss in Bell, p. xvii, next reference.

33. E. T. Bell, *Men of Mathematics* (Simon & Schuster, 1937), p. xvii.

34. E. A. Burtt, *Op. Cit.*, p. 53.

35. E. A. Burtt, *Op. Cit.*, p. 68.

36. Morris Kline, *Mathematics and the Physical World* (Dover, 1959), p. 385.

37. Morris Kline, *The Loss of Certainty* (Oxford University, 1980), p. 345, quoting from Jeans, *The Mysterious Universe.*

38. Hermann Weyl, *Philosophy of Mathematics and Natural Science* (Atheneun, 1963), p. 125, "...the question of the reality of the world mingles inseparably with the question of the reason for its lawful mathematical harmony... The ultimate answer lies beyond all knowledge, in God alone, emanating from him, the light of consciousness, its own origin hidden from it, grasps itself in self-penetration, divided and suspended between subject and object, between meaning and being".

39. Quoted by Morris Kline from *The Assayer* (1610) in *The Loss of Certainty, Op. Cit.*, p. 46.

40. Quoted by Morris Kline from Einstein's *The World as I See It* (1934) in *The Loss of Certainty, Op. Cit.*, p. 345.

41. In *Mathematics Today*, ed. L. A. Steen (Vintage, 1978).

42. Quoted by Morris Kline in *The Loss of Certainty, Op. Cit.*, p. 298.

43. Albert Einstein, *Essays in Physics* (Philosophical Library, 1930), p. 56.

44. Stephen W. Hawking, *A Brief History of Time: From the Big Bang to Black Holes* (Bantam, 1988).

45. Stephen W. Hawking, *Ibid.* Chap. 5, *Elementary Particles and the Forces of Nature*, pp. 63–80.

46. "It is more important to have beauty in one's equations than to have them fit experiment", P. A. M. Dirac, *The Evolution of the Physicist's Picture of Nature, Sci. Am.* **208** (5, 1963) pp. 45–53.

47. Ilkka Niiniluoto, *Varieties of Realism* in *Symposium on the Foundations of Modern Physics: 1987*, eds. Pekka Lahti and Peter Mittelstaedt (World Scientific, 1987) pp. 459–484.

48. *Scientific Realism*, ed. Jarrett Leplin (University of California, 1984).

49. *Images of Science*, eds. Paul M. Churchland and Clifford A. Hooker (University of Chicago, 1985).

50. Hilary Putnam, *What is Realism?* in Jarrett Leplin, *Op. Cit.*, pp. 140–153.

51. Richard N. Boyd in Leplin, *Op. Cit.*, p. 50.

52. Boyd in Leplin, *Op. Cit.*, p. 56.

53. Boyd in Leplin, *Op. Cit.*, p. 60.

54. Boyd in Leplin, *Op. Cit.*, p. 61.

55. Victor F. Weisskopf, *Knowledge and Wonder* (MIT, 1979), 2nd edition.

56. Weisskopf, *Ibid.* p. 129.

57. Weisskopf, *Ibid.* p. 138.

58. Nick Herbert, *Quantum Reality* (Anchor, 1985), Chap. 5, pp. 71–92.

59. This property of indefiniteness is captured in the so-called Heisenberg Uncertainty principle. The term "uncertainty" is misleading in that it suggests human vacillation. Actually, this principle is about the indefinite nature of quantum wave states with respect to some physical quantities. There is as much certainty about this feature of nature as there is for any other principle of physics. See Herbert, *Ibid.*, p. 109 for more on Heisenberg's principle.
60. Robert Hanbury Brown, *The Intensity Interferometer* (Halsted, 1974).
61. Herbert, *Op. Cit.*, Chaps. 11 and 12, pp. 199–232.
62. Bernard d'Espagnat, *In Search of Reality* (Springer-Verlag, 1979), Chap. 4.
63. J. C. Polkinghorne, *The Quantum World* (Longman, 1984), Chap. 7.
64. Fritz Rohrlich, *Facing Quantum Mechanical Reality, Science* **221** (1983) 1251.
65. Abner Shimony, *The Reality of the Quantum World, Sci. Am.* **249** (1987) 46.
66. Albert Einstein, *Essays in Physics* (Philosophical Library, 1950) p. 63. The title of the essay is *Fundamentals of Theorectical Physics*, originally published in *Science*, 24 May 1940.
67. Niels Bohr, *Discussion with Einstein on Epistemological Problems in Atomic Physics* in *Albert Einstein: Philosopher-Scientist*, ed. P. A. Schilpp (Open Court, 1949), pp. 199–242.
68. Niels Bohr, *Essays 1958/1962 on Atomic Physics and Human Knowledge* (Wiley, 1963), p. 60.
69. C. F. von Weizsäcker, *The Copenhagen Interpretation* in *Quantum Theory and Beyond*, ed. Ted Bastin (Cambridge University, 1971).
70. E. P. Wigner, *The Problem of Measurement* in *Symmetries and Reflections, Op. Cit.*, p. 164.
71. John von Neumann, *Mathematical Foundations of Quantum Mechanics* (Princeton University, 1955).
72. E. P. Wigner, *Remarks on the Mind-Body Question* in *Symmetries and Reflections, Op. Cit.*, pp. 175–176. See also p. 186.
73. David Bohm, *Hidden Variables and the Implicate Order* in B. J. Hiley and F. David Peat, *Quantum Implications* (Routledge and Kegan Paul, 1987),pp. 33–45.
74. J. P. Vigier *et al.*, *Causal Particle Trajectories and the Interpretation of Quantum Mechanics*, *Ibid.*, pp. 169–204.
75. Hans Primas, *Contextual Quantum Objects and Their Ontie Interpretation* in Pekka Lahti and Peter Mittelstaedt, *Symposium on the Foundations of Modern Physics* (World Scientific, 1987), pp. 251–276.
76. Roger Penrose, *Quantum Physics and Conscious Thought* in Hiley and Peat, *Op. Cit.*, pp. 105–120.
77. In *The Many Worlds Interpretation of Quantum Mechanics*, eds. B. S. DeWitt and R. D. Graham (Princeton University, 1973).

A Few Systems-Colored Views of the World

Yi Lin
Department of Mathematics
Slippery Rock University
Slippery Rock, PA 16057–1326
USA

1. What is Mathematics?

Mathematics is not only a rigorously developed scientific theory, but also the fashion center of modern science. That is because a theory will not be considered as a real scientific theory if the theory has not been written in the language and symbols of mathematics, since such a theory does not have the capacity to predict the future. Indeed, no matter where we direct our field of vision, whether to the depths of outer-space where we are looking for our "brothers" somewhere faraway, or to the slides beneath the microscopes where we are trying to catch the basic bricks of the world, or to the realm of imagination where infinitely many human ignorances are comforted, etc., we can always find the track and the language of mathematics.

In the long, long human history, many diligent plowers of the great scientific garden have cultivated today's magnificant mathematical foliage. Mathematics is not like any other theories in history, which have become faded again and again as time goes on, and for which new foundations and new fashions have had to be designed by coming generations in order to mesh new problems. History and reality have repeatedly showed that the mathematical foliage is always young and flourishing.

Mathematical abstraction has frightened many people. But at the same time, many unrecognized truths contained in mathematics have led truth-pursuers to be astonished by the sagacity of mathematics. In fact, mathematics is a splendid abstractionism with the capacity to describe, to investigate and to predict the world. First of all, let us see how some of the greatest thinkers in history have enjoyed the beauty of mathematics!

Philosopher Bertrand Russell[1]: "Mathematics, rightly viewed, possesses not only truth, but supreme beauty – a beauty cold and austere, like that of sculpture, without appeal to any part of our weaker nature, without the gorgeous trappings of painting and music, yet sublimely pure, and capable of a true perfection such as only the greatest art can show. The true spirit of delight, the exaltation, the sense of being more than man, which is the touchstone of the highest excellence, is to be found in mathematics as surely as in poetry. What is best in mathematics deserves not merely to be learned as a task, but to be assimilated as a part of daily thought, and brought again and again before the mind with ever-renewed encouragement. Real life is, to most men, a long second-best, a perpetual compromise between the real and the possible; but the world of pure reason knows no compromise, no practical limitations, no barrier to the creative activity embodying in splendid edifices the passionate aspiration after the perfect from which all great work springs. Remote from human passions, remote even from the pitiful facts of nature, the generations have gradually created an ordered cosmos, where pure thought can dwell as in its natural home, and where one, at least, of our nobler impulses can escape from the dreary exile of the natural world."

Mathematician Henri Poincaré[1]: "it (mathematics) ought to incite the philosopher to search into the notions of number, space, and time; and, above all, adepts find in mathematics delights analogous to those that painting and music give. They admire the delicate harmony of number and forms; they are amazed when a new discovery discloses for them an unlooked for perspective; and the joy they thus experience, has it not the esthetic character although the senses take no part in it? Only the privileged few are called to enjoy it fully, it is true; but is it not the same with all the noblest arts?"

Organically stacked colors show us the beautiful landscape of nature, so that we more ardently love the land on which we were brought up. Animate combinations of musical notes sometimes bring us to somewhere between the sky with rolling black clouds and the ocean with roaring tides, and sometimes comfort us in a fragrant bouquet of flowers with singing birds. How does mathematics depict the world for us? Mathematics has neither color nor sound, but with its special methods – numbers and forms – mathematics has been showing us the structure of the surrounding world. It is because mathematics is the combination of the marrow of human thoughts and the aesthetics that it can be developed generation after generation.

Let us see, for example, the concept of numbers. What is "2"? Consider two books and two apples. The objects, book and apple, are completely different. Nevertheless, the meaning of the concept of "2" is the same. Notice that the symbol "2" manifests the same inherent law in the different things – quantity. We know that there are the same amount of books and apples, because by pairing a book and an apple together, we get two pairs. Otherwise, for example, there are two apples and three books. After forming two pairs, each of which contains exactly one apple and one book, there must be a book left. From this fact an ordering relation between the concepts of "2" and "3" can be expected.

Colors can describe figures of nature, and musical notes can express fantasies from the depths of the human mind. In the previous example, every one will agree that the symbol "2" displays a figure of nature and insinuates the marrow of human thoughts. By using intuitive pictures and logical deduction, it can be shown that the totality of even natural numbers and the totality of odd natural numbers have the same amount of elements. That is because any even natural number can be denoted by $2n$, and any odd natural number can be described by $2n + 1$, where n is a natural number. Thus, for each natural number n, $2n$ and $2n+1$ can be put together to form a pair. With the help of this kind of reasoning and abstraction, the domain of human thoughts is enlarged from "finitude" into "infinitude", just as the human investigation of nature goes from daily life into the microcosm and cosmic space.

How can colors be used to display atomic structures? How can musical notes be used to imitate the association of stars in the universe? With mathematical symbols and abstract logic, the relationship between the stars in the universe and the attraction between atomic nucleus and electrons are written in detail in many research books.

Maybe you, the reader, would ask: Because mathematics, according to the previous description, is so universally powerful, can you use mathematics to study human society, human relations, and communication and transportation between cities?

The answer to this question is "yes". Even though the motivation for studying these kinds of problems appeared in Aristotle's time, the research along this line began not too long ago. Systems theory, one of the theories concerning these problems, was formally named in the second decade of this century, and the theoretical foundation began to be set up in the sixth decade. It is still not known whether the theory can eventually depict and answer the afore-mentioned problems. However, the theory has been used to explore some problems man has been studying since the beginning of human history, such as the problems: (1) Does there exist eternal truth in the universe? (2) Does the world consist of basic particles, which cannot be divided into smaller particles? Systems Theorists have eliminated the traditional research method – dividing the object under consideration into basic parts and processes, and studying each of them – and have begun to utilize systems methods to study problems with many cause-effect chains. For instance, the famous three-body problem of the sun, the moon, and the earth, the combination of vital basic particles, and social systems of human beings are examples of such problems.

Mathematics is analogous neither to painting, with its awesome blazing scenes of color, nor to music, which possesses intoxicatingly melodious sounds. It resembles poetry, which is infiltrated with exclamations and admirations about nature, a thirst for knowledge of the future, and the pursuit of ideals. If you have time to sit down to taste the flavor of mathematics meticulously, you will be attracted by the delicately arranged symbols to somewhere in the depths of mathematics. Even though gentle breezes have never carried any soft mathematical songs, man has been firmly believing that mathematics shows a truth since the first day it appeared. The following two little poems by Shakespeare[1] will give mathematics a better shape in your mind.

1

Music and poesy used to quicken you :
The mathematics, and the metaphysics,
Fall to them as you find your stomach serves you.
No profit grows, where is no pleasure ta'en : –
In brief, sir, study what you most affect.

2

I do present you with a man of mine,
Cunning in music and in mathematics,
To instruct her fully in those sciences,
Whereof, I know, she is not ignorant.

2. The Construction of Mathematics

Analogous to paintings, which are based on a few colors, and to music, in which the most beautiful symphonies in the world can be written with a few basic notes, the mathematical world has been constructed on a few basic names, axioms and the logical language.

The logical language K used in mathematics can be defined as follows. K is a class of formulae, where a formula is an expression (or sentence) which may contain (free) variables. Then K is defined by induction in the following manner, for details see Ref. 2 or 7:

(a) Expressions of the forms given below belong to K:

$$x \text{ is a set },x \in y ,x = y$$

as well as all expressions differing from them by a choice of variables, where the concept of "set" and the relation "$\in$" of membership are the primitive notations which are the only undefined primitives in the mathematics developed in this section.

(b) If ϕ and ψ belong to the class K, then so do the expressions (ϕ or ψ), (ϕ and ψ), (if ϕ, then ψ), (ϕ and ψ are equivalent) and (not ϕ).

(c) If ϕ belongs to K and v is any variable, then the expressions (for any v, ϕ) and (there exists v such that ϕ) belong to K.

(d) Every element of K arises by a finite application of rules (a), (b) and (c).

The formulae in (a) are called atomic formulae. A well-known set of axioms, on which the whole classical mathematics can be built and which is called ZFC axiom system, contains the following axioms:

I. *Axiom of extensionality.* If the sets A and B have the same elements, then they are equal or identical.

II. *Axiom of the empty set.* There exists a set $\emptyset$ such that no x is an element of $\emptyset$.

III. *Axiom of unions.* Let A be a set of sets. There exists a set S such that x is an element of S if and only if x is an element of some set belonging to A. The set S is denoted by $\cup A, X \cup Y$, if $A = \{X, Y\}$, or $\cup_{i=1}^{n} X_i$, if $A = \{X_1, X_2, \ldots, X_n\}$.

IV. *Axiom of power sets.* For every set A there exists a set of sets, denoted by $p(A)$, which contains exactly all the subsets of the set A.

V. *Axiom of infinity.* There exists a set of sets A satisfying the conditions: $\emptyset \in A$; if $X \in A$, then there exists an element $Y \in A$ such that Y consists exactly of all the elements of X and the set X itself.

VI. *Axiom of choice.* For every set A of disjoint non-empty sets, there exists a set B which has exactly one element in common with each set belonging to A.

VII. *Axiom of replacement for the formula ϕ.* If for every x there exists exactly one y such that $\phi(x, y)$ holds, then for every set A there exists a set B which contains those and only those elements y for which the condition $\phi(x, y)$ holds for some $x \in A$.

VIII. *Axiom of regularity.* If A is a non-empty set of sets, then there exists a set X such that $X \in A$ and $X \cap A = \emptyset$, where $X \cap A$ means the common part of the sets X and A.

The following two axioms can be deduced from Axioms I–VIII.

II'. *Axiom of pairs.* For arbitrary a and b, there exists a set X which contains only a and b; the set X is denoted by $X = \{a, b\}$.

VI'. *Axiom of subsets for the formula ϕ.* For any set A there exists a set which contains the elements of A satisfying the formula ϕ and which contains no other elements.

From Axioms II and IV, we can construct a sequence $\{a_n\}_{n=0}^{\infty}$ of sets as follows:

$$\begin{aligned}
&a_0 = \emptyset\,, \\
&a_1 = p(\emptyset) = \{\emptyset\} = \{a_0\}\,, \\
&a_2 = p(a_1) = \{\emptyset, \{\emptyset\}\} = \{a_0, a_1\}\,, \\
&\dots \\
&a_n = p(a_{n-1}) = \{a_0, a_1, \dots, a_{n-1}\}\,, \\
&\dots\,.
\end{aligned}$$

Now, it can be readily checked that the sequence $\{a_n\}_{n=0}^{\infty}$ satisfies the following Peano's axioms:

P1. 0 is a number.

P2. The successor of any number is a number.

P3. No two numbers have the same successor.

P4. 0 is not the successor of any number.

P5. If P is a property such that

(a) 0 has the property P; and

(b) whenever a number n has the property P, then the successor of n also has the property P, then every number has the property P.

In the proof, we understand the symbol 0 in P1 as $a_0 = \emptyset$, the notations of "number" in P2 as any element in the sequence $\{a_n\}_{n=0}^{\infty}$ and of "successor" in P2 as the power set of the number.

The entire arithmetic of natural numbers can be derived from the Peano's axiom system. The last axiom P5 embodies the principle of mathematical induction and illustrates in a very obvious manner the enforcement of a mathematical "truth" by stipulation. The construction of elementary arithmetic on Peano's axiom system begins with the definition of the various natural numbers, that is the construction of the sequence $\{a_n\}_{n=0}^{\infty}$. Because of P3 (in combination with P5), none of the elements in the sequence $\{a_n\}_{n=1}^{\infty}$ will be led back to one of the numbers previously defined, and in virtue of P4, it does not lead back to 0 either.

As the next step, a definition of addition can be established, which expresses in a precise form the idea that the addition of any natural number to some given number may be considered as a repeated addition of 1; the latter operation is readily expressible by means of the successor relation (that is, power set relation). This definition of addition goes as follows:

A1. (a) $n + 0 = n$; (b) $n + p(k) = p(n + k)$.

The two stipulations of this inductive definition completely determine the sum of any two natural numbers.

Now, the multiplication of natural numbers may be defined by means of the following recursive definition, which expresses in a rigorous form the idea that a product nk of two natural numbers may be considered as the sum of k terms each of which equals n.

A2. (a) $n \cdot 0 = 0$; (b) $n \cdot p(k) = nk + n$.

Within the Peano system of arithmetic, its true propositions flow not merely from the definition of the concepts involved, but also from the axioms that govern these various concepts. If we call the axioms and definitions of an axiomatized theory the "stipulations" concerning the concepts of that theory, then we may now say that the propositions of the arithmetic of the natural numbers are true by virtue of the stipulations which have been laid down initially for the arithmetic concepts.

In terms of addition and multiplication, the inverse operations, i.e., subtraction and division, can then be defined. But it turns out that these cannot always be performed; i.e., in contradistinction to the sum and the multiplication, the subtraction and the division are not defined for every pair of numbers; for example, 7–10 and 7÷10 are undefined. This incompleteness of the definitions of the subtraction and the division suggests an enlargement of the number system by introducing negative and rational numbers.

Consider the Cartesian product $N \times N = \{(n, m) : n, m \in N\}$, where N is the collection of all natural numbers defined as above, which is a set by Axiom V, and (n, m) is the ordered pair of the natural numbers n and m, which is defined by $(n, m) = \{\{n\}, \{n, m\}\}$. Consider a subset Z of the power set $p(N \times N)$ such that for any two ordered pairs $(a, b), (c, d) \in N \times N$, (a, b) and (c, d) belong to the same set A in Z if and only if $a + d = b + c$. Then it can be seen that each ordered pair in $N \times N$ is contained in exactly one element belonging to Z and no two elements in Z have an element in common. Now, the set Z is the desired set on which the definition of subtraction of integers can be defined for each pair of elements from Z, where the set Z is called a set of all integers, and each element in Z is called an integer. In fact, for each ordered pair $(n, m) \in N \times N$, there exists exactly one element $A \in Z$ containing the pair (n, m). So the element A can be denoted by $A = \overline{(n, m)}$. Then the addition, multiplication and subtraction can be defined as follows:

A3. $\overline{(a, b)} + \overline{(c, d)} = \overline{(a + c, b + d)}$;

A4. $\overline{(a, b)} \cdot \overline{(c, d)} = \overline{(ac + bd, ad + bc)}$;

A5. $\overline{(a, b)} - \overline{(c, d)} = \overline{(a + d, b + c)}$;

for any elements $\overline{(a, b)}$ and $\overline{(c, d)}$ in Z. It is important to note that the subset $Z^+ = \{\overline{(a, b)} \in Z$: there exists a number $x \in N$ such that $b + x = a\}$ of the set Z now serves as the set of natural numbers, because Z^+ satisfies axioms P1–P5.

It is another remarkable fact that the rational numbers can be obtained from the ZFC primitives by the honest toil of constructing explicit definitions for them, without introducing any new postulates or assumptions. Similarly, rational numbers are defined as classes of ordered pairs of integers from Z. The various arithmetical operations can then be defined with reference to these new types of numbers, and the validity of all the arithmetical laws governing these operations can be proved by virtue of nothing more than Peano's axioms and the definitions of various arithmetical concepts involved.

The much broader system thus obtained is still incomplete in the sense that not every number in it has a square root, cube root,..., and more generally, not every algebraic equation whose coefficients are all numbers of the system has a solution in the system. This suggests further expansions of the number system by introducing real and finally complex numbers. Again, this enormous extension can be effected by only definitions, without posing a single new axiom. On the basis thus obtained, the various arithmetical and algebraic operations can be defined for the numbers of the new system, the concepts of function, of limit, of derivatives and integral can be introduced, and the familiar theorems pertaining to these concepts can be proved, so that finally the huge system of mathematics as here delimited rests on the narrow basis of ZFC axioms: Every concept of mathematics can be defined by means of the two primitives in ZFC, and every proposition of mathematics can be deduced from ZFC axioms enriched by the definitions of the non-primitive terms. These deductions can be carried out, in most cases, by means of nothing more than the principles of formal logic – a remarkable achievement in systematizing the content of mathematics and clarifying the foundations of its validity.

Remark: A well-known result discovered by K. Gödel shows that the afore-described mathematics is an incomplete theory in the following sense. Even though all those propositions in the classical mathematics can indeed be derived, in the sense characterized above, from ZFC, there are other propositions which can be expressed in pure ZFC language and which are true, but which cannot be derived from ZFC axioms. This fact does not, however, affect the result outlined above. Because the most unreasonably effective part of mathematics in applications is a substructure of the mathematics constructed above.

3. Mathematics from the Viewpoint of Systems

The concept of systems was introduced formally by L. von Bertalanffy in the 1920's according to the following understanding about the world: The world we live in is not a pile of uncountably many isolated "parts"; and any practical problem and natural phenomenon cannot be described perfectly by only one cause-effect chain. The basic character of the world is its organization and the connection between the interior and exterior of different things. The customary investigation of the isolated parts and processes cannot provide a complete explanation of the vital phenomena. This kind of investigation gives us no information about the coordination of parts and processes. Thus the chief task of modern science should be a systematic study of the world. For detail see Ref. 4.

Mathematically speaking, S is said to be a system[3] if and only if S is an ordered pair (M, R) of sets, where R is a set of some relations defined on the set M. Any element in M is called an object of the system S, and the sets M and R are called the object set and relation set of the system S, respectively. In this definition, for any relation $r \in R$, the relation r is defined as follows: there exists an ordinal number $n = n(r)$, called the length of the relation r, such that r is a subset of M^n, where M^n indicates the Cartesian product of n copies of the set M. Assume that the length of the empty relation $\emptyset$ is 0; i.e., $n = n(\emptyset) = 0$. In ordinary language, a system consists of a set of objects and a collection of relations between the objects.

Claim 1. If by a formal language we mean a language which does not contain sentences with grammar mistakes, then any formal language can be described as a system.

Proof. Assume that L is a formal language; for the sake of convenience, suppose that L is the English which does not contain sentences with grammar mistakes, where an English grammar book B is chosen and rules in B are used as the measure to see if an English sentence is in L or not.

Each word in L consists of a finite combination of letters, that is, a finite sequence of letters. Let X be the collection of all twenty six letters in L. From the finiteness of the collection X and Axiom VII, it follows that X is a set; otherwise the natural number 26 is a set (from the discussion in the previous section), so instead of the collection X we can use the set 26. Let M be the totality of all finite sequences of elements from X, then M is the union below:

$$M = \bigcup_{i=1}^{\infty} X^i .$$

So M is a set, which contains all words in L.

Each sentence in the language L consists of a finite combination of some elements in M; i.e., a finite sequence of elements from M. Let Q be the totality of finite sequences of elements from M, then Q is the following union:

$$Q = \bigcup_{i=1}^{\infty} M^i .$$

Thus Q is a set. Each element in Q is called a sentence of the language L.

Write M as a union of finitely many subsets: $M = \cup\{M(i) : i \in J\}$, where $J = \{0, 1, \dots, n\}$ is a finite index set. Elements in $M(0)$ are called nouns; elements in $M(1)$ verbs; elements in $M(2)$ adjectives, etc.... Let $K \subset Q$ be the collection of all sentences in the grammar book B. Then K must be a finite collection and it follows from Axiom VII that K is a set.

If each statement in K is assumed to be true, then the ordered pair (M, K) constitutes the English L. Here, the pair (M, K) can be seen as a systems description of the formal language L. Q.E.D.

Let (M, K) be the systems representation of the formal language L in the above proof, and T the collection of Axioms I–VI′ in the previous section. Then $(M, T \cup K)$ can be seen as a system description of mathematics.

From the definition of systems given by G. Klir[5] that a system is what is distinguished as a system and Claim 1, it follows that if a system $S = (M, R)$ is given, then any relation $r \in R$ can be understood as an S-truth, i.e., the relation r is true among the objects in the set M. Therefore, any mathematical truth is a $(M, T \cup K)$-truth, i.e., it is derivable from ZFC axioms, the principles of formal logic and definitions of some non-primitive terms. From this discussion about mathematical truths the following question is natural:

Question 1. If there existed two mathematical statements derivable from ZFC axioms with contradictory meanings, would they still be $(M, T \cup K)$-truths? Generally, what is the meaning of a system with contradictory relations, e.g., $\{x < y, x \geq y\}$ as the relation set?

Another natural epistemological problem was asked in Ref. 6 as follows:

Question 2. How can we know whether or not there exist contradictory relations in a given system?

From Klir's definition of systems it follows that generally Questions 1 and 2 cannot be answered. About this end, a Gödel's theorem shows that it is impossible to show whether the systems description $(M, T \cup K)$ of mathematics is consistent (that is, any two statements derivable from ZFC axioms will not have contradictory meanings) or not in ZFC axiom system. On the other hand, we still have not found any method outside ZFC axiom system which can be used to show that the system $(M, T \cup K)$ is inconsistent. This means that there are systems, for example, the mathematics based on ZFC, so that we do not know whether there exist propositions with contradictory meanings. Therefore, this fact implies that maybe not every system is consistent or that not every system has no contradictory relations.

The application of mathematics has showed us that mathematics is extremely effective in describing, solving and predicting practical phenomena, problems and future events, respectively. That means that in practice a subsystem of the system $(M, T \cup K)$ can always be found to match a situation under consideration. For example, a hibiscus flower has five petals. The mathematical word "five" provides a certain description of the flower. This description serves to distinguish it from flowers with three, four, six,... petals. When a watch chain is suspended from its ends, it assumes very nearly the shape of the mathematical curve known as the catenary. The equation of the catenary is $y = \frac{a}{2}(e^{x/a} + e^{-x/a})$, where a is a constant. At the same time, this equation can be used to predict (or say, answer) the following problem: Suppose that someone holds his ten-inch watch chain by its ends and his fingers are at the same height and are four inches apart, how far below his fingers will the chain dip? Now, the following amazing question can be asked (note that this question was not originally posed by me):

Question 3. Why is mathematics so "unreasonably effective" as applied to the analysis of natural systems?

In order to discuss this question, we better go back to Axioms I–VI′. As a consequence of the discussion in this section, the whole structure of mathematics might be said to be true by virtue of mere definitions (namely, of the non-primitive mathematical terms) provided that ZFC axioms are true. However, strictly speaking, we cannot, at this juncture, refer to ZFC axioms as propositions which are either true or false, for they contain free primitive terms, "set" and the relation of membership "$\in$", which have not been assigned any specific meanings. All we can assert so far is that any specific interpretation of the primitives which satisfies the axioms – i.e., turns them into true statements – will also satisfy all the theorems deduced from them. For detailed discussion, see Refs. 2 and 7. But, the partial structure of mathematics developed on the basis of Peano's axioms has several – indeed, infinitely many – interpretations which will turn Peano's axioms, which contain the primitives: "0", "number", and "successor", into true statements, and therefore, satisfy all the theorems deduced from them. (Note that this partial structure of mathematics constitutes the theoretical foundation for almost all successful applications of mathematics.) For example, let us understand by 0 the origin of a half line, by the successor of a point on that half-line the point 1 inch behind it, counting from the origin, and by a number any point which is either the origin or can be reached from it by a finite succession of steps each of which leads from one point to its successor. It can then be readily seen that all Peano's axioms as well as the ensuing theorems turn into true propositions, although the interpretation given to the primitives is certainly not any of those given in the previous section. More generally, it can be shown that every progression of elements of any kind provides a true interpretation of the Peano's axiom system. This example illustrates that mathematics permits of many different interpretations, in everyday life as well as in the investigation of laws of nature and from each different interpretation, we understand something more about nature; and at the same time, because of this fact, we feel that mathematics is so unreasonably effective as applied to the study of natural problems.

In the following, three examples will be given to show how abstract mathematical structures developed on the basis of ZFC axioms but outside Peano's axiom system can also be used to describe and to study problems in materials science, epistemology and sociology.

4. A Description of the States of Materials

For an arbitrarily chosen experimental material X, the material X can be described as a system $S_x = (M_x, R_x)$, where M_x is the set of all molecules in the material and R_x contains the relations between the molecules in the set M_x, which describe the spatial structure in which the molecules in the material are arranged. If the material X is crystalline, then all cells in X have the same molecular spatial structure. That means that the molecules in X are aligned spatially in a periodic order, so that we have the following (all basic notations of set theory appearing in the following sections can be found in Ref. 2):

Definition 1. The experimental material X is crystalline, if for the system $S_x = (M_x, R_x)$, there exists a relation $f \in R_x$ such that $f|M^* \neq f$, for any proper subset M^* of M_x, and there exists a cover $\mathcal{B}$ for the set M_x such that for any A and B in $\mathcal{B}$, there exists a one-to-one correspondence $h : A \to B$ satisfying $f|A = f|B \circ h$, where $f|Y$ indicates the restriction of the relation f on the subset Y and $h(x_0, x_1, \ldots, x_\alpha, \ldots) = (h(x_0), h(x_1), \ldots, h(x_\alpha), \ldots), \alpha < \beta$, for any ordinal number β and any $(x_0, x_1, \ldots, x_\alpha, \ldots) \in A^\beta$.

For example, if the material X is crystalline, for any unit cell Y in X, define M_Y as the set of all molecules contained in the cell Y. Then $\mathcal{B}_x = \{M_Y : Y \text{ is a unit cell in } X\}$ is a cover of M. Because each molecule in X must belong to one unit cell; and all unit cells look the same. It follows from this fact that the main characteristics of crystalline materials are described by Definition 1.

Given a linearly ordered set W, a time system S over the set W is a function from W into a family of systems; say $S(w) = S_w = (M_w, R_w)$, for each $w \in W$, where each system S_w is called the state of the time system S at the moment w. If, in addition, a family $\{l_w^{w'} : w, w' \in W, w' \geq w\}$ of mappings is given, such that for any r, s and $t \in W$ satisfying $s \geq r \geq t, l_r^s : M_s \to M_r$ and

$$l_t^s = l_t^r \circ l_r^s \text{ and } l_t^t = \mathrm{id}_{M_t} ,$$

where id_{M_t} is the identity mapping on the set M_t, then the time system S is called a linked time system and denoted by $\{S_w, l_t^s, W\}$. Each mapping l_t^s is called a linkage mapping of the linked time system. For details, see Ref. 8.

Let R be the set of all real numbers, R_T $(=R)$ indicate the temperature and $R_t = [0, +\infty)$ and $R_p = [0, +\infty)$, the interval of real numbers from 0 to the positive infinity, indicate time and pressure, respectively. Define an "order" relation "$<$" on the Cartesian product $W = R_t \times R_T \times R_p$ as follows: for any $(x_1, y_1, z_1), (x_2, y_2, z_2) \in W, (x_1, y_1, z_1) < (x_2, y_2, z_2)$ if and only if either (i) $z_2 > z_1$, (ii) $z_1 = z_2$ and $y_1 > y_2$, or (iii) $z_1 = z_2, y_1 = y_2$ and $x_1 < x_2$. Then it can be shown that "$<$" is a linear order.

Now, let us establish a system theory model of crystallization process of the material X. It is assumed that the size of molecules in X will not change as temperature and pressure change; i.e., any chemical reaction between the molecules during the process will be neglected.

In the state of time t, temperature T and pressure P, the relation set, consisting of all relations describing spatial structure of the molecules in X, is F_w, where $w = (t, T, P)$. Then at the state w, the material X can be described as a system (M, F_w), where M is the set of all molecules in X, so that a linked time system S is obtained as follows:

$$\{(M, F_w), \mathrm{id}_M, W\} ,$$

where $\mathrm{id}_M : M \to M$ is the identity mapping on the set M.

If the material X is going to crystallize as time goes on, pressure imposed on X goes up and temperature goes down, then there is a state $w_0 = (t_0, T_0, P_0) \in W$ such that the material X is crystalline. That means that the state $S_{w_0} = (M, F_{w_0})$ of the linked time system S satisfies the property described in Definition 1.

The crystallization process of the material X is the process that X changes from an amorphous state to an ordered state or a crystalline state. The process is divided into two steps. The first one is called nucleation; i.e., when temperature is lower than the melting temperature of the material, the activity of the molecules decreases, which means that the attraction or interaction between molecules increases, so that there would be a tendency for the molecules to get together to form a short-range order. This kind of locally ordered region is called nucleus. The second step is called growth. Although at certain temperature and pressure values, many nuclei can be formed, at the same time, they will probably be disintegrated because the ratio of the surface and the volume of each of the nuclei is still too great and the surface energy is too high. However, as soon as the nuclei grow large enough, which is caused by the fluctuation of local energy, their energy will be lower as more and more molecules join them. The molecules in the nearby region would prefer to join those nuclei so that they will grow larger and larger.

Question 4. In the previous mathematical model, does there exist a relation $f \in R_x$ satisfying the property in Definition 1, when the material X is actually in a melting state, where $S_x = (M_x, R_x)$ is the systems-representation of the material X?

We assume that the answer to Question 4 is "yes". This assumption is based on the following:

(i) a view of dialectical materialism – the internal movement of contradictions dominates the development of the matter under consideration;

(ii) the beauty of the systems theory model described above.

Under this assumption, it can be seen that under certain ideal conditions any material may crystallize, because the relation f will be observable under certain ideal conditions. Suppose that Y is a cell in the crystalline material X. The spatial geometric relation (called the structure of the cell Y) of the molecules in the cell Y is termed the basic structure of the material X.

Generalizing the model in Ref. 9, the following is our model for the arbitrarily fixed material X: no matter what state the material X is in, such as in solid state, in liquid state or in gas state, there exist many pairwise disjoint groups of molecules in X such that each molecule in X is contained in exactly one of the groups, and every group is "topologically" isomorphic to the basic structure of the material X. The basic structure of the material determines all the physical properties the material X has. The meaning of "topologically isomorphic" was explained in Ref. 9 with examples.

Using this model, many phenomena about materials can be explained naturally. For details, see Ref. 9. In Ref. 10, the model was used to study the soil erosion by waterdrops.

5. Some Epistemological Problems

In ZFC axiom system, the following results can be proven:

(1) There does not exist a system whose object set consists of all systems.

(2) A system $S_n = (M_n, R_n)$ is said to be an nth-level object system of a system $S_0 = (M_0, R_0)$, if there exist systems $S_i = (M_i, R_i), i = 1, 2, \ldots, n-1$, such that $S_i \in M_{i-1}, i = 1, 2, \ldots, n$. Then for any system S, there does not exist any nth-level object system S_n of S such that $S_n = S$, for any natural number $n > 0$.

(3) A sequence $\{S_i\}_{i \in n}$ of systems, where n is an ordinal number, is said to be a chain of object systems of a given system S, provided S_0 is an i_0th-level object system of S, and for any $i, j \in n$, if $i < j$, then there exists a natural number $i_j > 0$ such that S_j is an i_jth-level object system of S_i. Then for any fixed system S, any chain of object systems of S must be finite.

In Ref. 6, the following epistemological problems were discussed: (i) The feasibility of the definition of the theory so-called "science of science". (ii) The existence of basic particles in the world. (iii) The existence of absolute truths.

The theory of science of science appeared in the 1930's. Since then more and more scientists have been involved in the research of the theory. The background in which the theory appeared is that: since the human society entered the twentieth century, the development of different aspects of science and technology have been sped up. The development of science shows a developing tendency to divide science into more and more disciplines and at the same time to synthesize results and methods in different fields together to get new results and understandings about the world. Hence, a natural problem arose: Can science be studied as a social phenomenon, so that man can develop scientific research with purposes and improve its recognition of the natural world? This problem is very important to administrators of scientific research, because so far any scientific achievement is obtained by either individual scientists or small groups of scientists. Thus there are many scientific achievements obtained independently by many different scientists. If the answer to the afore-mentioned problem is "yes", then it partially means that we can save at least some of the valuable scientific labors from doing the same things by arranging them to do some pre-targeted researches. In this way, the development of science will be sped up greatly. From the historical background mentioned above, some scientists think that the research object of the theory of science of science is the following: The theory of science of science is such a theory that instead of any real matter it studies science as a whole, the history and the present situation of each discipline, the relationship between disciplines and the whole developing tendency of science is studied.

Combining Claim 1 and result (2), it can be seen that the theory of science of science defined above cannot exist. Roughly speaking, it is because the theory has to contain the study about itself. For detailed discussion, see Ref. 6.

From the definition of systems given by Klir[5] and result (1) it follows that we cannot consider a relation which is true in a system that contains all systems as its objects. Does this imply that there is no universal truth? I believe so. I think that for any given truth, an environmental system is given, in which the truth is a true relation.

The result (3) above says that any system is built upon objects which are no longer systems. Does this imply that the world consists of basic particles, where a basic particle is a particle which cannot be divided into smaller particles? See Ref. 6 for details.

6. A Mathematical Discussion of a Social Phenomenon

In Ref. 11, the following social phenomenon was discussed mathematically in detail: in a relationship between the people in a group, one or a few of the people must dominate over the others, why? The concept of centralized systems was used to analyze the natural phenomenon.

Suppose that $A = (M_A, F_A)$ and $B = (M_B, F_B)$ are systems, A is said to be a partial system of B[3], if either (i) $M_A = M_B$ and $F_A \subseteq F_B$, or (ii) $M_A \subsetneqq M_B$ and there exists a subset F' of F_B such that $F_A = \{f|M_A\colon f$ is a relation in $F'\}$, where $f|M_A$ indicates the restriction of the relation f on the set M_A. A system $S = (M, F)$ is said to be a centralized system, if any object in M is a system and there exists a system $C = (M_c, F_c)$ with $M_c \neq \emptyset$ such that for any distinct elements x and $y \in M$, say $x = (M_x, F_x)$ and $y = (M_y, F_y)$, $M_c = M_x \cap M_y$ and $F_c \subseteq F_y|M_c \cap F_x|M_c$, where $F_y|M_c = \{f|M_c : f \in F_y\}$ and $F_x|M_c = \{f|M_c : f \in F_x\}$. The system C is called a center of the system S.

A set X is said to be uncountable, provided that X is an infinite set and there does not exist a one-to-one correspondence between the set X and the set N of all natural numbers. Then the following theorem was proved: Suppose that A is a system with an uncountable object set, and each object in A is a system with a finite object set. If there exists an element which is an object in at least uncountably many objects in the system A, then there exists a partial system B of A with an uncountable object set, and B forms a centralized system.

Suppose that A is a collection of people, and $[A]^{<\omega}$ is the collection of all finite subcollections in A. Then for any finite subcollection $x \in [A]^{<\omega}$, the following possibilities exist: (a) there exists exactly one relation between the people in x; (b) there exist more than one relation between the people in x; (c) there does not exist any relation between the people in x. If the situation (a) occurs, then we construct a system $S_x = (x, F_x)$ such that the relation set F_x contains only one relation which describes the relation between the people in x. If the situation (b) occurs, then we construct a collection $\{S_x^i : i \in I_x\}$ of systems, where I_x is an index set, such that for any $i \in I_x$, the relation set of the system S_x^i contains only one relation which describes a relation between the people in x; and for any fixed relation f between the people in x, there exists exactly one $i \in I_x$ such that $S_x^i = (x, \{f\})$. If the situation (c) occurs, we construct a system $S_x = (x, \emptyset)$, where the relation set of the system S_x is empty. Whence a collection M of systems has been defined:

$$\begin{aligned} M = \{S_x : x \in [A]^{<\omega}, &\text{there is at most one relation} \\ &\text{between the people in } x\} \cup \{S_x^i : i \in I_x, x \in [A]^{<\omega}, \\ &\text{there are more than one relation between} \\ &\text{the people in } x\}, \end{aligned}$$

and each system in M has a finite object set.

Now we consider the system $(M, \emptyset)$ with empty relation set. Then the theorem above tells that if the following conditions (1) and (2) hold, then there exists a subcollection $M^* \subset M$ such that $(M^*, \emptyset)$ forms a centralized system and the object set M^* is uncountable:

(1) The collection M is uncountable; and
(2) There exists at least one person who is an object in at least uncountably many elements in M.

From the construction of the system $(M, \emptyset)$ it follows that the condition (1) means that there exists a complicated network of relations between the people in A; and the condition (2) implies that there exists at least one person in A who has enough relations with the others in A. Therefore, it follows that no matter how complicated the relations between the people in a real community can be, the theorem says that if a person or a subcollection of people $x \in [A]^{<\omega}$ want to be a center in the collection A, then he or they must have the following background:

(3) the people in x have many transverse and longitudinal relations with the others in A.

Thus, because in any social society, there are uncountably many relations between the people in the society, from the point of view of systems we can now understand why there always exist factions in human society. (It is because there exist more than one center in human society.)

7. Final Comments and Further Questions

From the discussion above it can be seen that there roughly are two methods to introduce new mathematical concepts. One is that in the study of practical problems, new mathematical concepts are abstracted, and the other is that in the search for relations between mathematical concepts, new concepts are discovered. The application of mathematics, as shown by the three examples above, brings life to the development of mathematics so that when dealing with real problems, mathematics often makes us understand the phenomena under consideration better. If all well-developed mathematical theories do not fit a specific situation under consideration, a modified mathematical theory will be developed to study the situation. That is to say, mathematics, just as all other scientific branches, is developed in the process of examining, verifying and modifying itself.

When facing with practical problems, new abstract mathematical theories are developed to investigate the problems. At the same time, the predictions based on the new developed theories are so accurate that the following question appears:

Question 5. Does the human way of thinking have the same structure as that of the material world?

On the other hand, the mathematical palace is built upon "empty set $\emptyset$". And history gradually shows that almost all natural phenomena can be described and studied with mathematics. From these facts the following question arises:

Question 6. If all laws of nature can be written in the language of mathematics, can we conclude that the universe we live in rest on the "empty set $\emptyset$", just as what it is said in Holy Bible that "in the beginning, God created the heaven and the earth. And the earth was without form, and void"?

Acknowledgment The author wants to appreciate Ms. Teri Foster for her enthusiastic help in the preparation of the manuscript.

References

1. R. E. Moritz, *On Mathematics and Mathematicians* (Dover, 1942).
2. K. Kuratowski and A. Mostowski, *Set Theory: with an Introduction to Descriptive Set Theory* (North-Holland, 1976).
3. Y. Lin, *A Model of General Systems, Mathl. Modelling* **9** (1987) 95–104.
4. L. von Bertalanffy, *Modern Theories of Development*, translated by J. H. Woodge (Oxford University Press, 1934).
5. G. J. Klir, *Architecture of Systems Problem Solving* (Plenum, 1985).
6. Y. Lin, *A Multi-Relation Approach of General Systems and Tests of Applications. Synthese: An International J. for Epistemology, Methodology and Philosophy of Science* **79** (1989) 473–488.
7. K. Kunen, *Set Theory: An Introduction to Independence Proofs* (North-Holland, 1980).
8. Y.-H. Ma and Y. Lin, *Some Properties of Linked Time Systems, Int. J. General Systems* **13** (1987) 125–134.
9. Y. Lin and Y.-P. Qiu, *Systems Analysis of the Morphology of Polymers, Mathl. Modelling* **9** (1987) 493–498.
10. L. C. Wu and Y. Lin, *Systems Analysis of the Formation of Crators by Waterdrop Impact Throught Water Layer*, submitted for publication.
11. Y. Lin, *An Application of Systems Analysis in Sociology. Cybernetics and Systems: An Int. J.* **19** (1988) 267–278.

The Reasonable Effectiveness of Mathematical Reasoning

Saunders Mac Lane
Department of Mathematics
University of Chicago
Chicago, IL 60637, USA

That famous quotation from Eugene Wigner – on the unreasonable effectiveness of mathematics – came at a time when the standard attitude towards the nature of mathematics was primarily formal and hence misleading. Indeed, the view then dominant took mathematics to be strictly formal, partly because Bertrand Russell had opined that mathematics was not knowledge of the truth, but just correct deduction of conclusions from premises, true or not. Actually, the work of David Hilbert had much more influence, both because he and his followers made extensive use of axiomatic methods which seem to emphasize deductions from premise (axioms) and because of his foundational efforts. In order to achieve his desired consistency proof for all of mathematics he needed to treat mathematical proofs as systematic manipulation of formal symbols, so that he could argue (in what he called metamathematics) about these proofs as themselves objects of thought, in the hope of showing that no proper proof could ever reach a contradiction.

These two examples exhibit the doctrine of formalism: That mathematics is to be understood as a formal manipulation of sequences of symbols, with each mathematical theorem or lemma stated in strictly formal language and each proof a sequence of formal changes in accord with rules of inference specified in advance. Clearly a complete acceptance of the idea that mathematics is formal in just this sense meant that the practical uses of mathematics were inexplicable. Thanks to formalism in the foundations and the extensive development of axiomatic methods (Abstract algebra, functional analysis and the like), it would indeed seem unreasonable to expect that mathematics could be so effective as it in fact was and is.

Now mathematics does have this formal aspect. However, ultimately the forms that arise here are all based on and derived from earlier experience with the facts. It is then for this reason that mathematical form will be effective in describing the facts, both the originally noted ones and new facts. Moreover our experience shows that many of the facts exhibit steady patterns which exemplify form.

This article will go on to describe a different analysis of the nature of mathematics which can account for its utility. The crux of the matter is that mathematics is an integral part of exact scientific knowledge, and that its inevitable formal aspect is tightly bound up with its role in the development of such exact knowledge.

First, mathematics is indeed in one aspect something "formal"; indeed, one might "define" mathematics by saying that it is that part of scientific knowledge of the world which can be reduced to formulae and their objective manipulation. Thus to take a simple example, we do not need to add 3 cows to 4 cows to get 7 cows or \$3 to \$4 to get \$7, we can just add 3 to 4 to get 7, because the result will always be the same, 7, whether we are counting cows or cash – and this formal rule for addition is effective because it works both for cows and for cash – as well as other things. The laws of ballistics, formulated with differential equations, allow one to calculate where a projectile will go from a given start, and the equations are the same, whether the projectile is a baseball, a bullet, or a rocket. Thus the formal does not have content because it fits several different contents. Mathematics is there because there are in the world very many contents which have the same form. In particular, calculus can be formal because it arose from the common aspects of different phenomena in astronomy and mechanics.

This also applies to arithmetic. The familiar rules of arithmetic have been developed from repeated observations and long experience with counting and listing discrete things. And the experimentally based rules for adding and multiplying numbers can be presented formally – just learn the multiplication table and how to check your result, by casting out 9's or otherwise. Thus 5 times 7 means that a rectangular area 5 feet wide and 7 feet long will be 35 square feet and that 5 pens at \$7 each will cost \$35, and that a 7-hour hike at 5 miles per hour will cover 35 miles. The fact that the formal rules of multiplication are so effective is not due to chance or some other mystery but due to the careful choice of the rules on the base of experience – in this case the observation that multiplication is repeated addition and that the same rules of addition apply to many things: In this case, the experience that area, cost and distance could all be treated in a systematic way, and that these systematic ways were all the same when the different operations were reduced to the standard manipulation of numbers by formal rules.

Once the rules are found, they work for very large numbers as well. It then appears that some numbers are not products of smaller numbers and so are primes – and that there are an infinity of such primes. Moreover, every whole number can be written as a product of primes in a unique way. So theorems about numbers appear. In particular, new objects such as the primes fascinate. Thus the facts about the distribution of the primes in the sequence of whole numbers lead to studies having no direct relation to the original use of numbers for counting. But in the last analysis, these primes did arise out of experience – so it is not really remarkable when they find use in the practical manipulation of codes and ciphers. What is remarkable is the depth of the properties of primes, and their relation to calculus, as in the analysis of the famous Riemann Zeta function, which uses the full array of properties of the calculus for complex numbers to display the distribution of the primes.

To summarize, arithmetic starts with counting and discrete aspects of the world which are then detached from their origins, give rise to number theory with properties developed abstractly, detached from practical considerations. But since these properties arose in the first place from practice, it is by no means surprising that they can then be profitably used with other aspects of reality.

Many other parts of mathematics have similar and sometimes hidden practical origins: Calculus started from mechanics, Fourier series from acoustics and thermodynamics, and complex analysis from optics, thermodynamics, and integration, while probability theory had its origins in thinking about gambling; its later uses are much more widespread.

The case of geometry is similar. Long human experience with shapes and the constructions of buildings lead to empirical observations about properties of shapes which could be formulated in terms of angles and triangles – for example to the property of a right angle stated in the pythagorean theorem. Gradually, other such uniformities are recognized and then formalized, and further investigation reveals that the results can be organized in a systematic way, starting with selected evident facts as axioms from which the other known results can be deduced as theorems. Thus for example, the pythagorean theorem on right triangles can be deduced either from properties of similar triangles or from additive properties of areas. In this way, the axiomatic method was born from the formal organization of real world experience with shapes. Once discovered, this method has many other successes in organizing other formal properties extracted from experience.

In the real world, most shapes are not triangular or rectilinear, but curved. Thus formal geometry is led to consider circles and spheres. Then other curves such as ellipses and parabolas exhibit curvature, but a curvature which varies from point to point. At any one point, this curvature can be measured by approximating the variable curve by circles, and this approximation then makes use of the techniques of approximation developed in the calculus (for different purposes); the result is a calculus formula for curvature – forms found in one place again apply elsewhere. Next, the contemplation of various curved surfaces leads to the measurement of curvature in different directions at a point on a surface, and thence to the measure of the Gaussian curvature which is presently a part of Riemannian geometry. Thus form begets new form. For a period, it seemed that the development of Riemannian geometry and the related tensor analysis in the late 19th century was just a part of pure mathematics – but then it was used in the description of space-time, in general relativity theory. Thus it is that tensor analysis is remarkably effective in physics, but this effect is not unreasonable, because the geometry here had earlier been extracted from geometrical aspects of the world.

Today, this remarkable interaction between geometry and physics has continued in new directions. Thus geometers had inevitably to consider tangents and other things attached to points of a curved space, and so bundles over such a space, described with suitable tensors and differential forms. This led to the definition of a connection in such a fiber bundle; it turned out to be essentially the same thing as a gauge field in physics. This remarkable confluence of ideas had led to new discoveries both in physics and mathematics – as for example in the discovery by Michael Freedman and Simon Donaldson that there can be more than one smooth structure on a Euclidean space of four dimensions (something which cannot happen in dimensions other than 4). This example, and many others, indicates that the apparently purely formal structures in geometry include items which are precisely those needed to formulate the geometric aspects of the real world. In brief, abstract geometry can apply because it originally came from experience with geometric shapes and geometric extent.

The mathematical study of symmetry exhibits the same type of connection between the formal and the real. Symmetry was present in many things, for example in Greek architecture, and some aspects of symmetry were present in geometry from its first beginnings, as in the study of the regular solids. The full mathematical description of symmetry was developed only slowly, with many preliminaries. First, a symmetry f of an object can be described as an imagined motion of the object to a new position which exactly covers the original position – as when a regular octagon is rotated by an angle of $2\pi/8$ about its center. If f and g are two such symmetries of the same object it follows that the composite motion fg – f followed by g – is also a symmetry. This explains the basic relevance of composition to symmetry. There were many examples of such symmetries and their composed symmetries, and it was centuries before the notion was abstracted to that of a symmetry group of an object – the collection of all the symmetries $f, g \dots$ and their compositions. Thus a transformation group T of an object A may be described as a set of symmetries $f, g : A \to A$ which contains the inverse f^{-1} of each symmetry, the identity symmetry, and the composite $f \circ g$ of any two symmetries. This composite is necessarily associative, in that $f(gh) = (fg)h$. Then a final abstraction gives the notion of an (abstract) group: A collection G of elements $x, y, z \dots$ (or $f, g, h \dots$) such that any two have a composite xy in G, where this composition is associative, has an identity and to each x an inverse x^{-1} such that $x^{-1}x = 1 = xx^{-1}$. This notion of a group is a prime example of abstract mathematical form – defined in logical terms which do not require any reference to examples. But it was first by examples that the form arose. Actually, it was first explicit in an algebraic equation where the problem was to find, if possible, algebraic solutions to 5th degree equations like the solution for quadratics and cubic equations. Then it was Galois who studied the group of permutations of the roots of such equations – as in the interchange of the two roots i and $-i$ of the equation $x^2 + 1 = 0$. It then turned out that the nature of the group of symmetries for an equation of degree 5 meant that there could not be the desired formula by radicals for its roots.

Groups also arose in more practical questions of crystallography, where studies by C. Jordan, Fedorov, and Schoenfliess eventually found the 230 crystallographic space groups (see the reference to the book by J. J. Burckhardt). Group theory also entered in the classification of the various sorts of geometry studied in the 19th century, as in the famous Erlanger Program (1872) of Felix Klein, which classified geometries by their groups of symmetry. This is typical for mathematical form: A notion, here that of a group, arises from a number of different examples – here architectural symmetry, crystal symmetry, algebraic symmetry of equations and symmetries of different geometries. This is why mathematics is inevitably formal in its expression: because it must apply to many different situations, locating the same form in each. Even in the mathematics itself, there is a tension between the concrete and the abstract – as in the study of abstract groups by means of their representations by groups of transformations – usually linear or orthogonal transformations of spaces. The result for groups is the striking confluence visible today, where pure group theory prospers, as in the recent classification of all possible finite simple groups and where the possible group representations are dominant tools in the description of elementary particles in physics. Given its origins, it is no wonder that group theory is effective.

Many other examples, not to be listed here, illustrate my basic thesis as to the nature of mathematical knowledge. The external world exhibits patterns which repeat themselves and so can be captured as forms. The same form can appear in different concrete guises, and so can effectively be described abstractly, without reference to any one concrete realization. Such an abstract consideration of the form as such makes it possible to deduce its various properties, independently of the different guises. The mathematics lies in these deductions, even when the investigations go far beyond the original appearance of the forms. Then combination of different forms leads to new mathematical objects, which in some cases can then be used to understand facts of the physical or social world. This is why mathematics is effective: the world exhibits regularities which can be described, independently of the world, by forms which can be studied and then reapplied.

This view of the nature of mathematics, under the label "Formal functionalism", has been described in more detail in my recent book "Mathematics, Form and Function" in brief, mathematics is effective because it analyses the forms extracted from the functioning of the world.

References

1. J. J. Burckhardt, *Die Symmetrie der Kristalle* (Birkhauser Verlag, 1988) 196 pp.
2. Daniel S. Freed and Karen K. Uhlenbeck, *Instantons and Four-Manifolds* (Springer-Verlag, 1984) 232 pp.
3. Arthur Jaffe, *Ordering the Universe: The role of Mathematics, Notices of the Am. Math. Soc.* **31** (1984) 589–608.
4. Saunders Mac Lane, *Mathematics, Form and Function:* (Springer-Verlag, 1986) 476 pp.
5. George W. Mackey, *Induced representations of groups and Quantum Mechanics* (Benjamin, 1968) 167 pp.
6. Hermann Weyl, *Symmetry* (Princeton University Press, 1959) 168 pp.

Three Aspects of the Effectiveness of Mathematics in Science*

Louis Narens and R. Duncan Luce
Irvine Research Unit in Mathematical Behavioral Science
University of California
Irvine, CA 92717, USA

*This paper was supported, in part, by National Science Foundation Grant IRI-8996149 to the University of California, Irvine.

1. Introduction

Wigner (1960), in a widely read and cited article, articulated what had previously been recognized by many scientists, namely, the remarkable affinity between the basic physical sciences and mathematics, and he noted that it is by no means obvious why this should be the case. The remarkableness of this fact is obscured by the historical co-evolution of physics and mathematics, which makes their marriage appear to be natural and foreordained. But serious philosophical explanations for the underlying reasons are not many.

This paper dissects the normal use of mathematics in theoretical science into three aspects: idealizations of the scientific domain to continuum mathematics, empirical realizations of that mathematical structure, and the use in scientific arguments of mathematical constructions that have no empirical realizations. The purpose of such a dissection is to try to isolate gaps in our knowledge about the *justified* use of mathematics in science. We believe that the apparent "unreasonableness" referred to in Wigner's title "The unreasonable effectiveness of mathematics in the natural sciences" arises from such lack of knowledge, rather than to some principled difference in the nature of mathematics and that of science.

Of the three aspects mentioned, the second – empirical realizations of mathematical structure – is the one with the largest body of positive results. A number of these are recent, and a major portion of the paper is devoted to outlining some of them.[a]

2. Infinite Idealizations

In the sciences, the domains of interest are usually finite. But, from a human perspective, such domains are often "large" and "complex" and typically, scientists idealize them to infinite ones – to something ontologically much larger and more complex than the original domain. Paradoxically, the resulting models are frequently much more manageable mathematically than the more realistic, finite models. Although such idealized domains are necessarily not accurate descriptions of the ones of actual interest, these infinite, ideal descriptions are nevertheless useful in science. In the authors'

[a]The reader who wishes to know more about this and related research in the theory of measurement is referred to two expository papers by the authors, Luce and Narens (1987) and Narens and Luce (1986), and to two sets of books: Krantz *et al.* (1971), Suppes *et al.* (1989), and Luce *et al.* (1990); and Narens (1985; in preparation).

view, this is because the focus of science is generally not on giving exact descriptions of the entire domain under consideration – such descriptions would generally prove to be too unwieldy and too complex to be of use – but rather with the production of generalizations that capture particular, scientifically significant features of the domain and that interrelate nicely with other generalizations about related domains.

Many kinds of generalizations used in science require us to ignore certain properties that are inherent in finite models; and others emphasize properties that are valid only in infinite ones. For example, in finite, ordered domains, maximal and minimal elements necessarily exist. Further, a desirable generalization may exclude such elements, as for example: "For each American middle class individual, there is a slightly richer American middle class individual." Or a useful generalization may use the concept of a homogeneous domain (as explained below) – a condition that is a basis of many scientific laws and one that usually requires infinite domains.

The crux of the matter is that science – at least as currently practiced – relies on special kinds of generalizations rather than exact descriptions, and such generalizations necessarily need infinite domains, and hence actual scientific domains need to be idealized to infinite ones. There are many ways to carry out such idealizations, and unfortunately at the present time there are, in our view, no acceptable formal theories of idealization. While this is obviously a very important problem in the foundations of science, it appears to be a very difficult one. Various strategies have been suggested – e.g., idealize through some use of potential infinity or through some recursive process – but so far these attempts have failed to capture the full power of the current practices of mathematical science.

Intuitively, the idealization of a finite domain should be to some denumerably infinite one. However, mathematical science routinely employs nondenumerable domains (e.g., continua, finite-dimensional Euclidean space, etc.) for idealizations, because these have special, desirable modeling properties that are not possible for denumerable domains. (E.g., for continua X the proposition, "Each continuous function from a closed interval of X into X takes on maximum and minimum values," is true, whereas for densely ordered, denumerable X it is false.) While it is an interesting exercise to try and replace assumptions that imply nondenumerability of particular classes of scientific models with others that are consistent with denumerability, we will not go into this topic here, since for this paper it is somewhat peripheral.

Instead, we will simply take as a starting point that science employs idealizations of actual (non-mathematical) domains, and that these qualitative idealizations, which are usually infinite (and often nondenumerable) are used qualitatively to express generalizations about the original domain. But we emphasize that this is a major lacuna in the philosophy of science, one about which we do not, at present, have anything useful to say. In our opinion, understanding exactly when such idealizations are helpful is one major gap in understanding the effectiveness of mathematics in science and is a part of its perceived "unreasonableness".

With such idealizations accepted, the next problem is then to explain *why mathematics is so effective in drawing useful inferences about the qualitative, idealized domain.* To investigate this, the properties of qualitative structures that permit their quantification will be looked at first; that is, we will first look at how numbers enter in science through processes of measurement.

3. Empirical Bases for Number Systems

The number systems most used in science are algebraic subsystems of the ordered field of real numbers,[b] $\langle \mathrm{Re}, \geq, +, \bullet \rangle$. There are many such subsystems, and for the purposes of this paper, the most important ones are the subsystems of positive integers, positive rationals, and positive reals, respectively $\langle I^+, \geq, +, \bullet \rangle$, $\langle Ra^+, \geq, +, \bullet \rangle$, and $\langle Re^+, \geq, +, \bullet \rangle$. The subsystem of positive integers has two important empirical realizations: a system based on ordered counting (ordinal numbers) and a system based upon matching finite sets in terms of their being in one-to-one correspondence (cardinal numbers). Although these two realizations are *empirically different*, they are *algebraically identical*, since each can be shown to be isomorphic to the algebraic system of positive integers.

The system of positive rationals has a variety of empirical realizations, some in terms of the realizations of the system of positive integers. These latter will not be discussed in this paper for lack of space. Instead, we will focus on those that result from extensive measurement (discussed below), which is basic to much of mathematical science and which can also produce empirical realizations of the system of positive real numbers.

[b] It has become conventional in part of the measurement literature to use *Re*, *Ra*, and *I* for the real numbers, the rationals, and the integers, respectively, and to superscript them with + to denote the positive restrictions.

In understanding the role of numbers in science, it is important to describe the *empirical correlates* to the number domain, to the numerical ordering relation $\geq$ on it, and to the operations $+$ and $\bullet$ on it. In our view, this is not accomplished successfully for continuous domains by the well-known classical mathematical construction of the nineteenth century due to Peano and Dedekind, which constructs the system of the positive rationals out of the positive integers, and then the system of positive real numbers out of the positive rational numbers. The reason for its failure is that the *empirical correlates* that one wants for addition and multiplication on the reals are *not related in any empirical way via the construction* to the empirical realization(s) of the positive integers.

The way we shall proceed in this paper is to start with algebraic structures in which the domain, ordering relation, and addition operation are empirically realizable, and then investigate various empirical ways to realize multiplication.

3.1. Extensive Measurement

Extensive structures are algebraic structures of the form $\mathcal{X} = \langle X, \succsim', \bigcirc \rangle$, where X is a non-empty set of objects; $\succsim'$ is a binary relation on X; and $\bigcirc$ is binary operation on X, called the *concatenation operation.* It is assumed that:

(i) $\succsim'$ is a total ordering;
(ii) $\bigcirc$ is associative and commutative;
(iii) $\bigcirc$ is monotonic in the sense that it is strictly increasing relative to $\succsim'$ in each variable;
(iv) $\bigcirc$ is positive ($x \bigcirc y >' x$), where $>'$ is the strict part of $\succsim'$; and
(v) the structure $\mathcal{X}$ is such that all elements are "commensurable" in the sense that no two distinct elements of X are infinitely far apart nor infinitesimally close together in terms of the ordering and concatenation operation. (Elements x and y, where $x >' y$, are said to be "infinitely far apart" if and only if for all finite positive integers n, n concatenations of y remains strictly less than x, where for example "three concatenations" of y is defined by $(y \bigcirc y) \bigcirc y$. "Infinitesimally close" has a similar but slightly more complicated formulation.)

The important theorem about extensive structures is that they are isomorphically imbeddable in the numerical structure $\langle Re^+, \geq, + \rangle$, and from this it follows immediately that empirical realizations of extensive structures are empirical realizations of subsystems of the positive additive reals.

Extensive structures have many empirical realizations, including many of the basic physical dimensions such as length, charge, mass, etc. For example, for physical length, concatenation is accomplished by placing two perfectly straight measuring rods end-to-end in a line and forming a third rod by abutting them; and $\succeq'$ is determined by putting two rods side-by-side in the same direction with left endpoints corresponding and observing which spans the other.

In empirical realizations, the extensive ordering is the empirical correlate to the usual numerical ordering, and the extensive concatenation operation is the empirical correlate to numerical addition. What is missing is an empirical correlate to numerical multiplication. In measurement theory, the empirical correlate to multiplication rarely is a directly observed additional concatenation operation; it usually appears in much more subtle and indirect ways.

Throughout the rest of the paper, we will assume, unless explicitly stated otherwise, that the extensive structures discussed are either mappable onto $\langle Ra^+, \geq, +\rangle$ or $\langle Re^+, \geq, +\rangle$. These are the two most important situations and ones that can easily be described in terms of the defining relations, $\succeq'$ and $\bigcirc$, of extensive structures.

We now consider three different ways in which multiplication can be empirically introduced.

3.2. One Empirical Base for Multiplication: Multiplicative Representations

The first way considers multiplication as transformed addition. Although this idea, in itself, will not suffice to explain multiplication in the system of positive reals, in particular its distributivity over addition, it is nevertheless instructive to examine it.

Extensive structures $\langle X, \succeq', \bigcirc\rangle$ have isomorphisms into $\langle Re^+, \geq, +\rangle$ and thus have isomorphisms into $\langle (1, \infty), \geq, \bullet\rangle$. By deleting positivity from the assumptions of an extensive structure, a generalization results that has isomorphisms into $\langle Re^+, \geq, \bullet\rangle$, and thus may have an identity element (i.e., an element mapping into 1) and negative elements (those mapping onto (0,1)). In science, such generalized extensive structures often appear in indirect ways, especially in those situations where an attribute is affected by two (or more) factors that can be manipulated independently. Examples abound. In physics, varying either the volume and/or the substance filling the volume affects the resulting ordering by mass. In psychology and economics, varying the amount of a reward and the delay in receiving it each affects its ordering according to value. For a detailed discussion, see Chap. 6 of Krantz *et al.* (1971).

Abstractly, there are two disjoint sets X and P and a relation $\succsim$ on $X \times P$, where $(x,p)\succsim(y,q)$ is interpreted to mean that (x,p) exhibits at least as much of the attribute in question as does (y,q). Four empirically testable properties are invoked:

(i) $\succsim$ is a *weak order*, i.e., $\succsim$ is *transitive* – if $(x,p)\succsim(y,q)$ and $(y,q)\succsim(z,r)$, then $(x,p)\succsim(z,r)$ – and *connected* – either $(x,p)\succsim(y,q)$ or $(y,q)\succsim(x,p)$.

(ii) *Independence*, which amounts to monotonicity in each factor separately, i.e., if $(x,p)\succsim(y,p)$ for some p, then $(x,q)\succsim(y,q)$ for all $q \in P$, and a similar statement for the P-component.

From these two properties, it follows that there is a unique weak order induced on each factor, which is denoted $\succsim_i$, $i = X, P$, namely $x \succsim_X y$ if $(x,p)\succsim(y,p)$ for some p, and a similar statement for $\succsim_P$.

(iii) Each $\succsim_i$ is a total order.

The final property is a cancellation one that involves two equivalences implying a third.

(iv) *Thomsen Condition*: if $(x,r) \sim (z,q)$ and $(z,p) \sim (y,r)$, then $(x,p) \sim (y,q)$.

As in the case of an operation, two less testable conditions are also invoked. One is a form of solvability which in its strongest version asserts that given any three of $x, y \in X, p, q \in P$, the fourth exists such that $(x,y) \sim (p,q)$. The other is an Archimedean property that guarantees that all elements are commensurable. From these, one then shows that an operation can be induced on a component, say $\bigcirc_X$ on X, that captures all of the information about the conjoint structure, and that $\langle X, \succsim_X, \bigcirc_X\rangle$ is a generalized extensive structure with an isomorphism into $\langle Re, \geq, +\rangle$. This representation can then be reflected back to give a representation of $\mathcal{C} = \langle X \times P, \succsim\rangle$, namely, there are two positive real functions ψ_i, $i = X, P$, such that their sum is order preserving, i.e.,

$$(x,p)\succsim(y,q) \text{ iff } \psi_X(x) + \psi_P(p) \geq \psi_X(y) + \psi_P(q) .$$

Such structures are called *additive conjoint* ones. By taking exponentials, the representation is transformed into a *multiplicative representation* (θ_X, θ_P), i.e.,

$$(x,p)\succsim(y,q) \text{ iff } \theta_X(x)\theta_P(p) \geq \theta_X(y)\theta_P(q) ,$$

where $\theta_i = \exp \psi_i, i = X, P$. There is, so far, no reason to favor one kind of representation over the other since we do not have distinct notions of addition and multiplication that should be related by the usual distribution property, $r(s+t) = rs + rt$. We turn to that approach next.

3.3. A Second Empirical Base for Multiplication: Automorphisms

Let $\mathcal{X} = \langle X, \succeq', \bigcirc \rangle$ be an extensive sturcture. Consider the class of transformations of X *onto* itself that leave $\succeq'$ and $\bigcirc$ invariant – the *automorphisms* of the structure. These, of course, capture symmetries of the structure in the sense that everything "looks the same" before and after the transformation. Although the concept of "automorphism" is highly abstract, individual automorphisms and sets of automorphisms are often realized empirically. For example, in the structure $\mathcal{X}$ it is easy to verify empirically that functions like $g(x) = x \bigcirc x$ and $h(x) = (x \bigcirc x) \bigcirc x$ are automorphisms.

A very important fact about the usual axiomatization discussed above is that the structure is *homogeneous* in the sense that each element "looks like" each other element: given any two elements, there is an automorphism that takes the one into the other. Another important fact is that it is *1-point unique* in the sense that if two automorphisms agree at one point, then they agree at all points. Both homogeneity and 1-point uniqueness, despite their abstractness, play an important role in the *empirical* aspects of measurement theory. This is because they are often implied by purely empirical considerations.

Let m be an isomorphism of $\mathcal{X}$ onto $\mathcal{R} = \langle R, \geq, + \rangle$, where R is either Ra^+ or Re^+. Let α denote an automorphism of $\mathcal{X}$, and for $x \in X$, let $m(x)$ denote the number assigned to it. Then a function f_α is defined on R by:

$$f_\alpha[m(x)] = m[\alpha(x)] .$$

Since the automorphisms form a group with many nice properties (see below), these functions combines nicely and define the operation we know as multiplication. Among other things, it is related to addition by the familiar law of distribution, which reflects nothing more than the fact that it arises from automorphisms: For suppose $x, y \in X$. Then

$$\begin{aligned} f_a[m(x) + m(y)] = f_\alpha[m(x \bigcirc y)] &= m[\alpha(x \bigcirc y)] \\ &= m[\alpha(x) \bigcirc \alpha(y)] \\ &= m[\alpha(x)] + m[\alpha(y)] \\ &= f_\alpha[m(x)] + f_\alpha[m(y)] . \end{aligned}$$

It is well-known that this equation defines multiplication in the usual sense, i.e., we may write for some positive *real* r_α,

$$f_\alpha[m(x)] = r_\alpha m(x)\,.$$

In this notation, the automorphism property becomes simply

$$r(s+t) = rs + st\,,$$

which is referred to as the *distribution* of multiplication over addition.

It should also be noted that one can show that changing the element to which the numeral 1 is assigned induces an automorphism of the structure, and so a change of unit is reflected in the numerical representation as multiplication by a constant.

3.4. A Third Empirical Basis for Multiplication: Distributive Operations

Consider an additive conjoint structure $\mathcal{C}$ that has an independently given operation $\bigcirc$ on the first component such that $\mathcal{X} = \langle X, \succsim_X, \bigcirc\rangle$ satisfies the properties of an extensive structure. We then say that the operation $\bigcirc$ *distributes in* $\mathcal{C}$ if and only if for each $x, y, u, v \in X, p, q \in P$, if $(x,p) \sim (y,q)$ and $(u,p) \sim (v,q)$, then $(x \bigcirc u, p) \sim (y \bigcirc v, q)$. This is just the sort of property that holds when X is a domain of volumes, P of substances, and $\succsim$ is an ordering by mass. Let ϕ_X be an additive representation of $\mathcal{X}$. Then one is able to show that we may take $\theta_X = \phi_X$ in a multiplicative representation (θ_X, θ_P) of $\mathcal{C}$. Thus, the main consequence is that relative to the operation $\bigcirc$, the impact of the second component P is that of an automorphism of $\mathcal{X}$, since multiplications are the automorphisms of the isomorphic image $\langle Re^+, \geq, +\rangle$ of $\mathcal{X}$. So by what we showed earlier, ordinary distribution of multiplication over addition again obtains. For details, see Luce and Narens (1985) and Narens (1985).

3.5. What Else is Measurable?

For many sciences, there do not appear to be many, if any, observable extensive structures on which to base measurement. To some degree the situation is alleviated by the representations of additive conjoint structures, but even that continues to place a very high premium on additivity – either associativity of an induced operation or equivalently, at the observational level, the Thomsen condition. Many concatenation and conjoint situations simply do not admit additive or multiplicative representations. Does this mean that measurement of such variables is impossible? Some, starting with Campbell (1920, 1928), have held that one must either have an extensive structure at the empirical level or, as some now accept, be able to show that an extensive structure is implicit, as in additive conjoint measurement. The question is whether more is possible at the implicit level, and the claim of modern work is, "Yes, a lot more."

The key new idea for this was developed in a series of papers (Alper, 1987; Cohen and Narens, 1979; Luce, 1986, 1987; Narens, 1981a,b) and is summarized in Chapter 20 of Luce, Krantz, Suppes, and Tversky (1990). The major mathematical result, which was completed by Alper (1987), is that for any ordered structure *on* the positive real numbers that is both homogeneous (see the definition given in the section on automorphisms) and finitely unique in the sense that for some integer N, any two automorphisms that agree at N distinct points are identical, there exists an isomorphic structure on the positive real numbers for which the automorphism group lies between the similarity group ($x \to rx, r > 0$) and the power group ($x \to rs^s, r > 0, s > 0$). An important consequence is that each member of this widely diverse class of structures has a subgroup of the automorphisms – called the *translations* – that is isomorphic to $\langle Re^+, \geq, + \rangle$. This consequence is useful because we can characterize readily what this means qualitatively.

In general, for an arbitrary, totally-ordered structure $\mathcal{X} = \langle X, \geq', R_j \rangle_{j \in J}$, where each R_j is a relation of finite order on X, define the *translations* to consist of the identity map together with all automorphisms α of $\mathcal{X}$ that do not have a fixed point (i.e., for all $x \in X, \alpha(x) \neq x$). The set T of translations can be ordered as follows: For $\sigma, \tau \in T, \sigma \geq^{''} \tau$ if $\forall x \in X, \sigma(x) \geq' \tau(x)$. The critical assumption is that $\mathcal{T} = \langle T, \geq^{''}, * \rangle$ is an Archimedean, totally-ordered group, where $*$ denotes function composition. If this is so, then $\mathcal{T}$ is isomorphic to a subgroup of $\langle Re^+, \geq, + \rangle$. This is essentially the same construction as for extensive structures. We will refer to $\mathcal{T}$ as the *implicit extensive structure* of $\mathcal{X}$.

If, further, $\mathcal{T}$ is homogeneous, then it is not difficult to show that $\mathcal{X}$ can be mapped isomorphically into the translation group, so that the translations of the isomorphic image of $\mathcal{X}$ are left multiplications of the translation group of $\mathcal{X}$. By the above, $\mathcal{X}$ is isomorphic to a numerical structure that has multiplication by positive real numbers (i.e., similarities) as its translations. Thus, measurement is effected for any ordered relational structure whose translations form a homogeneous, Archimedean ordered group. This class is far from vacuous: For example, consider any operation $\bigcirc$ on the positive reals, Re^+, for which there is a function $f : Re^+$ onto Re^+ such that f is strictly increasing, $f(x)/x$ is strictly decreasing, and $x \bigcirc y = yf(x/y)$. Then the structure $\langle Re^+, \geq, \bigcirc \rangle$ has as its translations the similarity group. The details can be found in Cohen and Narens (1979) and Luce and Narens (1985).

If this sort of implicit extensive measurement is deemed acceptable – and there is a growing body of literature which suggests that it is – then the study of measurement reduces to uncovering the properties of the set of translations. Doing so is not necessarily easy. An important problem is to give empirical conditions that for a broad class of situations will insure that the translations form a homogeneous group and reasonable conditions that will insure it is Archimedean. Some progress along these lines has been reported.

A question can be raised about these general structures that closely parallels the issue of the distribution of an extensive structure in a conjoint one.

3.6. *Distribution of General Relational Structures in Conjoint Ones*

Given our more general notion of measurement based upon an implicit extensive structure for the translations, one must inquire about the extent to which it behaves in familiar ways. In particular, it is important to understand the extent to which such general measures can be incorporated into the structure of physical quantities, which classically has involved just extensive structures distributing in conjoint ones. The first issue is to arrive at a suitable concept of distribution. This is done in two steps. First, given a conjoint structure $\mathcal{C} = \langle X \times P, \succsim \rangle$, two ordered n-tuples (x_i) and (y_i) from X are said to be *similar* if and only if there are $p, q \in P$ such that for $i = 1, \ldots, n$, $(x_i, p) \sim (y_i, q)$. Now, consider a relational structure on the first component of $\mathcal{C}$, namely, $\mathcal{X} = \langle X, \succsim_X, R_j \rangle_{j \in J}$, where each R_j is a relation of finite order on X. We say that $\mathcal{X}$ *distributes over* $\mathcal{C}$ if and only if for each $j \in J$, if (x_i) and (y_i) are similar ordered $n(j)$-tuples and (x_i) is in R_j, then (y_i) is also in R_j. A careful examination of the earlier definition of distribution for an extensive structure shows it to be a special case of the more general concept.

The key result (Luce, 1987) is that if the translations of a relational structure are homogeneous, form a group under function composition (or equivalently, are 1-point unique), and are Archimedean under the order defined earlier, then that structure is isomorphic to a relational structure on the first component of an additive conjoint structure and it distributes over the conjoint structure. Conversely, if an additive conjoint structure has a relational structure on the first component that distributes over the conjoint structure, then the translations of the relational structure form a homogeneous, Archimedean-ordered group.

The significance of this result is that to the extent we are interested in measurement that relates to conjoint structures via the distribution property, then the condition that the translations form a homogeneous, Archimedean-ordered group is exactly what is needed because it provides for the existence of the proper numerical setting in which addition and multiplication arise simultaneously with natural emprical correlates.

4. The Use of Non-Empirical Mathematics

We take the following generalization of the above remarks as one of our basis theses: *In many empirical situations considered in science – particularly in classical physics – there is a good deal of mathematical structure* already present *in the empirical situation. Measurement produces numerical correlates of that structure.*

We have seen several ways in which the algebraic system of positive real numbers can arise in science as numerical correlates of empirical concepts. By extending the methods, the system can be generalized to include powers and logarithms. More inventive methods could probably produce additional empirical correlates to simple concepts of integration and differentiation. Thus, it is reasonable to expect that important parts of elementary mathematics concerned with analysis have empirical realizations. If science only used such "elementary" means, then the scientific effectiveness of mathematics would be easily understood. However, science also uses additional sophisticated mathematical concepts and methods to get results, and many – if not most – of these methods have no empirical correlates. Symbolically, this situation can be expressed as follows:

An empirical situation E has an empirical mathematics E_M associated with it, that through a measurement representation, m, is isomorphic to a fragment M_E of mathematics, M. A scientist uses M, sometimes including portions of $M - M_E$, to get a result r about E. For this result to be 'about E', it must somehow be translatable into E. With what has so far been given, this can only be done if r is a result about M_E and is translatable into E via m^{-1}. In other words, M is used so that $m^{-1}(r)$

can be concluded about E_M, which in turn is used to draw the empirical conclusion r about E. If this use of M is "unreasonably effective", then its "unreasonableness" consists in using some part of $M - M_E$ to draw conclusions about M_E, a matter formulated purely in terms of mathematics, and one whose explanation we believe is likely to be found within mathematics itself.

Put another way, it is commonplace in science that non-interpretable mathematics – that is, mathematics with no empirical correlates – is used to draw mathematical conclusions that are empirically interpretable. In our opinion, such practices have not been adequately justified.

5. Conclusion

Our analysis has had three parts:

First, actual empirical situations are usually conceptualized as structures on large finite sets. Because of the complexity of such structures and the irregularities and non-homogeneity often necessarily inherent in them, the actual empirical situation is often idealized to an infinite "empirical" situation, where the irregularities and non-homogeneities disappear; that is, the actual situation is idealized to a more mathematically tractable structure. A well reasoned account of the conditions under which such idealizations are acceptable is a major unresolved problem in the philosophy of science.

Second, in many important areas of science, such idealized empirical structures contain – often in non-obvious ways – a good deal of mathematical structure, which through the process of measurement is realized as familiar mathematical structures on numerical domains. So it is not unreasonable to expect the kinds of mathematics inherent in such numerical realizations to be effective in producing conclusions that can be translated back to the idealized empirical situations (and thereby to the actual empirical situations) by inverting the measurement process. This aspect of the problem is relatively well developed and understood.

Third, and what remains a mystery, is why mathematics outside of the numerical realizations should be so (unreasonably) useful in determining correct results about those realizations. This appears to be more of a problem in the philosophy of mathematics than in the philosophy of science.

References

1. T. M. Alper, *A classification of all order-preserving homeomorphism groups of the reals that satisfy finite uniqueness*, *J. Math. Psychology* **31** (1987) 135–154.
2. N. R. Campbell, *Physics: The Elements* (Cambridge University Press, 1920) reprinted as *Foundations of Science: The Philosophy of Theory and Experiment* (Dover, 1957).
3. N. R. Campbell, *An Account of the Principles of Measurement and Calculation* (Longmans, 1928).
4. M. Cohen and L. Narens, *Fundamental unit structures: a theory of ratio scalability*, *J. Math. Psychology* **20** (1979) 193–232.
5. D. H. Krantz, R. D. Luce, P. Suppes and A. Tversky, *Foundations of Measurement* (Academic Press, 1971), vol. 1.
6. R. D. Luce, *Uniqueness and homogeneity of ordered relational structures*, *J. Math. Psychology* **30** (1986) 391–415.
7. R. D. Luce, *Measurement structures with Archimedean ordered translation groups*, *Order* **4** (1987) 165–189.
8. R. D. Luce, D. H. Krantz, P. Suppes and A. Tversky, *Foundations of Measurement*, (Academic Press, 1990), vol. III.
9. R. D. Luce and L. Narens, *Classification of concatenation measurement structures according to scale type*, *J. Math. Psychology* **29** (1985) 1–72.
10. R. D. Luce and L. Narens, *The mathematics underlying measurement on the continuum*, *Science* **236** (1987) 1527–1532.
11. L. Narens, *A general theory of ratio scalability with remarks about the measurement-theoretic concept of meaningfulness*, *Theory and Decision* **13** (1981a) 1–70.
12. L. Narens, *On the scales of measurement*, *J. Math. Psychology* **24** (1981b) 249–275.
13. L. Narens, *Abstract Measurement Theory* (MIT Press, 1985).
14. L. Narens (in preparation), *A Theory of Meaningfulness.*
15. L. Narens and R. D. Luce, *Measurement: The theory of numerical assignments*, *Psychological Bulletin* **99** (1986) 166–180.
16. P. Suppes, D. H. Krantz, R. D. Luce and A. Tversky, *Foundations of Measurement*, (Academic Press, 1989), vol. II.
17. E. P. Wigner, *The unreasonable effectivess of mathematics in the natural sciences*, *Commun. Pure Appl. Math.* **13** (1960) 1–14.

Mathematics and Natural Philosophy

Robert L. Oldershaw
15 West Pelham Road
Shutesbury, MA 01072, USA

1. Introduction

The unique and highly effective role that mathematics plays in the natural sciences has been an important topic in the realm of natural philosophy, and one that rarely fails to inspire epistemological debate. The background material that has been used in preparing this essay includes works by Wigner,[1] Hamming,[2] Kline,[3], Hardy,[4], Bronowski,[5] Weyl,[6] Singh,[7] and Einstein.[8] With notable exceptions, many of the previous expositions on the math-science relationship have been hampered by intrusions of obscurantism, anecdotal reasoning, and avoidance of definite conclusions. The present author, with all the naive confidence of a newcomer to this debate, herewith makes yet another attempt to provide a self-consistent set of answers to a series of age-old questions associated with the math-science relationship.

In Sec. 2 various terms and concepts are defined in an attempt to minimize the degree to which subsequent arguments are devalued by semantic "wooliness". Section 3 presents and defends four basic propositions about nature and our capacity to understand nature; these propositions constitute the core beliefs of the author's natural philosophy. In Sec. 4 the tools fashioned in the previous sections are applied to a battery of fundamental questions concerning the character of the math-science relationship.

The ideas and conclusions expressed in this essay are defended by logical argument and empirical example, but, as with virtually any issue in natural philosophy, there is ample room for disagreement. The author is not attempting to prove that his views are the only valid ones, but rather the author presents these candid ideas so that they might be considered by others who are interested in this subject matter and compared with their own viewpoints.

2. Definitions

One of the most basic requirements for a meaningful answer is that the terms and concepts used in the answer must be reasonably well-defined. The definitions presented below are ones that, in the author's judgement, are well-suited to an expression of his views on the particular issues of natural philosophy that are under consideration. They are not put forward as "THE" definitions.

(1) *Nature* Also referred to here as the physical world, the cosmos or the universe, the author uses the term nature to mean literally *everything*,

including all physical entities, their properties and their interactions with other entities.

(2) *Internal/External Universe* For the sake of the discussions below, the author uses the expression internal universe to refer to that exceeedingly limited portion of nature that exists or takes place within the brains of sentient beings. The external universe is defined as everything else. Dividing up the universe in this way is clearly artificial, but it is nonetheless useful in the following discussions.

(3) *Ideas* Any definition of the term idea (thought, concept, etc.) runs into the problem that one is dealing with an elemental thing that is hard to define in terms of other things, and the problem that the definition itself must involve ideas. Here the term idea will mean a reproducible, highly coordinated and poorly understood pattern of neuronal activity that represents information, that can give rise to other ideas within a brain, and that can be translated and transmitted to the external universe, and then to other parts of the internal universe, via entities or processes such as spoken language (sound waves) or graphic language (written or pictorial symbols). According to this definition, ideas only exist within the internal universe; there are no ghostly ideas flying around in the external universe, but only coded representations of ideas which have no meaning until they re-enter the internal universe and reach another brain.

(4) *Human/Absolute Knowledge* Human knowledge will be defined as ideas (factual, conceptual or analytical) that are constrained by human limitations in observing and thinking (processing ideas within the brain). Absolute knowledge will be defined as unconditionally true ideas based on absolutely perfect information.

(5) *Natural Science* The term natural science, or just science, is used to denote: (a) the existing human knowledge that pertains to nature and that has been tested in accordance with the "scientific method", and (b) the search for additional and/or superior human knowledge of nature.

(6) *Natural Philosophy* The author's *personal* definition of this term is: all natural science that is *not* applied mathematics, i.e., empirical knowledge (the vast amount of observational data), non-mathematical concepts (the Sun is a star, life evolves, etc.), and principles (the Copernican principle, the principle of relativity, conservation of mass/energy, etc.). Included within this large and diverse accumulation of scientific tools and knowledge are initial boundary conditions, the use of pattern recognition, symmetry

arguments, analogies, the scientific method, Occam's razor, concepts of determinism and causality, epistemological theories, etc. Mathematics is instrumental to the growth and refinement of natural philosophy, but it is the author's opinion that a given body of scientific knowledge can be divided into components of natural philosophy and applied mathematics.

(7) *Epistemology* This term is defined as the study of the means by which knowledge is acquired and the limits upon that acquisition process.

(8) *Mathematics* Again, we are confronted with something that is so elemental that it is very difficult to define. The definition of a thing can be based on more elemental or better known things, on the properties of the thing to be defined, or on concrete examples of the thing to be defined. The book *What is Mathematics?* by Courant and Robbins[9] uses the definition-by-examples method, while Hamming[2] tries the definition-by-properties method. The present author, wishing to steer as clear as possible from this quagmire, chooses the definition-by-examples method (with a vengeance), and defines mathematics as anything that professional mathematicians have recognized as mathematics. Forgive me if I do not list the members of this evolving set.

(9) *Abstract* When the author refers to something as being abstract, he means that it was created without the intention of directly or indirectly modelling something in nature. Obviously there can be a range of degrees of abstraction.

(10) *Language* This term is defined simply and broadly as the systematic use of symbolism as a means of communicating ideas.

3. Four Propositions about Nature and Knowledge

Before addressing important questions concerning the math-science relationship, the author would like to candidly identify some fundamental tenets of his natural philosophy and offer some supporting arguments for them.

Proposition 1: The universe is a cosmos. The term universe can refer to any kind of universe, but a cosmos is a universe that is highly ordered. We are part of a cosmos; one whose limitless complexity is evenly matched by an indescribably exquisite orderliness. I regard the validity of this proposition as virtually self-evident. For a glimpse of this notion of a balance between infinite complexity and elegant underlying order, consider the Mandelbrot set and its pictorial representations.[10] Here one sees limitless morphological variation, order and detail and yet, it is all produced by a short computer code for solving one exceedingly simple iterative equation and plotting its "behavior". Those who advocate the new "anthropic principle" do an admirable job of describing examples of the remarkable orderliness and the delicate balance of natural phenomena,[11,12] though I think that they then use these facts to support some rather dubious conclusions. My own conclusion regarding the extreme orderliness of the cosmos, and astronomical observations suggest that the same orderliness has prevailed for billions of years, is that it is due to the fact that all natural phenomena are governed by a set of fundamental laws of nature. My suspicion is that there is a relatively small number of these laws of nature, and that the orderliness and finely tuned balance of the cosmos are the necessary result of a universe governed by a finite set of elementary laws. Again, I would invoke the Mandelbrot set as an appropriate *metaphor* for nature's complexity, underlying order and elementary laws. Some have proposed that it is the human mind that has imposed order upon a universe that lacks inherent order. This position has always seemed to me to suffer from excessive hubris and from a tendency to ignore well-observed facts (e.g., the remarkable ordering of matter into atomic, stellar and galactic systems). At any rate, when I refer to the "laws of nature", I invoke the hypothesis that there are a finite number of elementary, universal and timeless laws that ultimately give rise to and strictly govern all natural phenomena. I should also state my bias in cosmology towards a universe that is without any spatio-temporal bounds, and so I reinterpret the "big bang" in terms of a "little bang" taking place in an infinite cosmos with a hierarchical organization.

Proposition 2: Pure abstraction is a myth It seems to me that human knowledge is inextricably linked to our experience of nature. As Albert Einstein stated it: "Pure logical thinking cannot yield us any knowledge of the empirical world; all knowledge of reality starts from experience and ends in it."[13] Certainly the advance of human knowledge of nature can be aided by *steps* that can be called pure logical thinking (Hamming[2] cites such a step by Maxwell), but in general most forms of human knowledge are directly derived from our experiences of nature. Even in fields wherein modelling of nature is not the goal, the most abstract thinking appears to be, at least ultimately, tethered to nature in two ways. Firstly, such abstract thinking invariably is the product of an evolution whose origins can be traced back to simple modelling of the world we experience. Secondly, many ideas and conceptual tools that are actively used in abstract thought are ideas that are directly derived from our experience of nature, such as the concept of number. To me, it seems sensible to acknowledge various degrees of *partial* abstraction, but wrong to believe that absolutely abstract thought is possible.

Proposition 3: Absolute knowledge of nature is forever beyond our reach Because our knowledge of nature is based on incomplete information, human knowledge of the cosmos (empirical, theoretical or whatever) cannot qualify as absolute knowledge. Our models of phenomena have always been approximations and, at best, we can only hope for asymptotic approach to the true facts and laws of nature. The lessons embodied in the recognition that nature's geometry is not strictly Euclidean, in the theorems of Godel, and in the new field of "chaotic" systems are all too clear. Even the simplest, most "self-evident" truths such as $1 + 1 = 2$ and the Earth orbits the Sun do not qualify as absolute truth (the former is not always true[2] and the latter is valid only for a finite time period). In fact, after much consideration, the author is willing to bet that he can find a flaw (such as conditional validity) in anything that anyone would like to propose as a candidate for a piece of absolute knowledge *of nature.* I suppose $C = 2\pi r$ qualifies as a piece of "absolute knowledge" within an ideal strictly Euclidean "world", but this is a far cry from unconditional absolute truth.

Proposition 4: Nature can only be the way it is One often encounters statements like: "If a particular fundamental constant were slightly different, then the world would be radically different." My *guess* is that there is only one infinite (spatially and temporally) cosmos, not a multitude of different parallel and non-interacting universes. Nature's fundamental laws, and properties such as constants, simply are what they are and could not have been otherwise.

4. Key Questions on the Math-Science Relationship

(1.) *Do Humans Invent Mathematics or Do They Discover Mathematical Truths that are Inherent in Nature?*

The most commonly cited exposition of the idea that humans "discover" mathematics is Hardy's statement: "I believe that mathematical reality lies outside of us, that our function is to discover or *observe* it, and that the theorems which we prove, and which we describe grandiloquently as our creations, are simply our notes of our observations. This view has been held, in one form or another, from Plato onwards... ."[4] The other school of thought on this matter, and apparently the more widely held view today, is put quite bluntly by Bridgeman: "It is the merest truism evident at once to unsophisticated observation, that mathematics is a human invention."[14] The present author shares the latter view that mathematics is obviously a human creation, but I also have considerable sympathy for Hardy's position because I feel that most of mathematics is directly or indirectly inspired (Proposition 2) by the order inherent in nature (Proposition 1). My synthesis of these opposing positions in the ancient "creation versus discovery" debate can be understood in terms of an analogy to an artist who has produced a sculpture of a horse. It is indeed self-evident that the artist created the sculpture, but it is also true that the artist was recording what he saw in the horse. I see the development of mathematics as involving *both* partially abstract invention and direct learning from the order of nature. In any particular instance one of these aspects might dominate, or they might play comparable roles, but assuredly mathematics, as a whole, *is* created by humans and *is* largely a reflection of our experience of nature. Thus it seems to me that the great "creation versus discovery" debate has been an unproductive affair with both sides taking unrealistically extreme positions, talking past each other, and insisting that the positions are mutually exclusive. According to Proposition 2, it is unacceptable to ignore the role that nature has played in the human creation of mathematics. On the other hand, according to Proposition 3, it is untenable to claim that our mathematical "discoveries" represent absolute knowledge, i.e., we do not "discover" the true order of nature, but rather we "discover" approximations to the natural order.

(2.) To What Extent is Mathematics Literally Abstract?

This question is closely related to its predecessor, and the author's answer can be readily anticipated from Proposition 2 and its discussion. The major branches of mathematics (such as number, geometry, calculus, etc.) arose directly from human experience of nature. That experience was both a source of inspiration and a vehicle for demonstrating the utility of mathematical creations. Obviously, mathematics has changed much since its empirical origins, but even the most soaring flights of abstraction are evolutionarily tethered to those empirical origins, and concepts directly derived from our experience of nature (e.g., numbers, geometric fundaments, sets, etc.) currently play an *active* role in mathematical reasoning.[7] Here one might profitably use the metaphor of mathematics being like a tree. Its roots are sunk into the earth of physical experience, and though its new growth rises ever farther from the earth, any part that is cut off from the "lifeblood" supplied by its roots would quickly dessicate and fall off. So the author maintains that absolute mathematical abstraction is a romantic myth, and that the partial abstraction of mathematics can range from minimally abstract to highly abstract. We cannot evade the physical world, and one can only wonder why one would want to be separated from its beauty, unless one is intimidated by its awesome complexity and *apparent* chaos. The "passions and follies" of mankind might cause anyone to seek refuge from the human "world", but no wise man seeks refuge from the intellectual experience of the physical world.

(3.) Is Mathematics a Science?

The author regards testability (or falsifiability) with respect to the order of nature as a *sine qua non* of science.[15] Accordingly the author does not ordinarily regard pure mathematics as a science. But by choosing a broader definition of the term science one can successfully defend including pure mathematics in that category. Therefore this question does not seem to be very useful because the term science is even less well-defined than the term mathematics.

(4.) Is Mathematics a Language?

Again we have a question that is largely a semantic one, but it does provide a vehicle for interesting comparisons of mathematics and spoken languages. Using the very general definition of the term language that is given in Sec. 2, mathematics certainly qualifies as a language, or a collection of languages. But let us look at two properties that mathematical language does not share with spoken languages. Firstly, as stressed by Margolis,[16] pure mathematical language is context-free. Unlike spoken languages wherein meaning can be dependent upon context (e.g., you have the right to turn right, especially if you support the political right), a crucial property of pure mathematics is that it remains independent of context, and therein lies its austere rigor. Secondly, one can undertake complex calculations with mathematics; try doing that with a spoken language! Still, when mathematics is applied to natural phenomena it often performs a descriptive function like that of a spoken language, albeit a uniquely succinct and quantitative description.

(5.) Is Mathematics the Language of Nature?

If this question is taken literally, then the author would have to answer firmly in the negative, since an affirmative answer would seriously contradict Proposition 3. An appropriate aphorism by Einstein[8] is: "as far as the propositions of mathematics refer to reality, they are not certain; and as far as they are certain, they do not refer to reality." Our mathematical creations are *approximations* to the actual language of nature whose true character is anybody's guess at this point. If I were to make a speculative guess as to the true character of nature's language (i.e., the elemental laws), then my candidate would be something like geometry, although infinitely more sophisticated than our current geometrical models.

(6.) Is Mathematics the Language of Science?

Yes, mathematics is the official, formal language of science but bear in mind that in the total absence of other forms of descriptive language (spoken or pictorial) it would be meaningless. Try it yourself: consider any non-trivial natural phenomenon and see how far you get in trying to describe (explain, analyze, etc.) it using only mathematics. How do you define the meaning of your mathematical symbols, how do you convey the initial conditions,...? The non-mathematical language of natural philosophy plays an equally important role in science.

(7.) How Effective is Mathematics as Applied to the Natural Sciences?

I think the answer to this question should be: very effective! Testimony to this high degree of effectiveness is ubiquitous in any science library and so it need not be repeated here. The reason for this effectiveness has to be that the extreme orderliness of the cosmos (Proposition 1) both inspires and validates mathematical modelling, which provides a glimpse at the actual underlying order, albeit through many veils. There are, however, limits to the effectiveness of applied mathematics and we would do well to never lose sight of that fact. Often we hear boastful assertions that a new mathematical application (e.g., string theory) has elucidated the true laws of nature and heralds a complete understanding of nature. We have heard it before and, I suppose, we will hear it again, and again,. . . .

(8.) Is the Effectiveness of Applied Mathematics "Unreasonable"?

Wigner[1] stated that: "the enormous usefulness of mathematics in the natural sciences is something bordering on the mysterious and. . . there is no rational explanation for it." He concluded the same essay with the remark: "The miracle of the appropriateness of the language of mathematics for the formulation of the laws of physics is a wonderful gift which we neither understand nor deserve." Hamming,[2] in a more down-to-earth essay, still ended up with a similar conclusion: "I am forced to conclude both that mathematics is unreasonably effective and that all of the explanations I have given when added together simply are not enough to explain what I set out to account for." But is it so unreasonable that human approximations (Proposition 3) of a very orderly cosmos (Proposition 1) should turn out to be very effective approximations? Returning to the sculptor analogy used above, is it "unreasonable" to expect that his sculpture of the horse bears a strong literal resemblance to the actual horse, or that it effectively captures impressions of the horse's movement or character? If the sculpture's "effectiveness" does not strike us as unreasonable, then why should the effectiveness of applied mathematics appear to be unreasonable? The author's answer is that it should not. Perhaps we sometimes romanticize and mysticize mathematics and mathematical physics in order to reassure ourselves that we are doing something quite extraordinary.

To the author it seems a little strange to claim, after very lengthy trial-and-error and/or logical efforts to find appropriate mathematical models of natural phenomena or mathematical ideas for solving problems in pure mathematics, that the *successful* results (and let's not forget the far more numerous examples of less happy results) of mathematical endeavors are "unreasonably" effective. In the next question, the author will address one of the more interesting arguments for the contention that mathematics is "unreasonably" effective.

(9.) Do We Get More Out of Mathemtics than We Put In?

Arguments that mathematical effectiveness is "unreasonable", "mysterious" and a "miracle" have often been largely based on the very loose notion that we get more out of mathematics than we put into it. The implication here is that mathematics has a "life of its own" and "is wiser than we are". But is this really the case? Firstly, let us look at some specific examples that are cited as the best evidence for this argument.

Hertz's discovery of the electromagnetic waves predicted by Maxwell's equations is perhaps the most common example. As Hamming[2] tells it: "to some extent for reasons of symmetry (Maxwell) put in a certain term, and in time the radio waves that the theory predicted were found by Hertz."

A second example, given by Wigner,[1] is the applicability of Heisenberg's matrix mechanics, "or a mathematically equivalent theory", to the helium atom for which the original assumptions should not have applied. Wigner calls this "the miracle of helium". Wigner[1] also cites the Lamb shift predicted by quantum electrodynamics as an example of how mathematics can go beyond our limited empirical knowledge. "The quantum theory of the Lamb shift, as conceived by Bethe and established by Schwinger, is a purely mathematical theory and the only direct contribution of experiment was to show the existence of a measurable effect." A final example might be the fact that QED correctly predicts the electron's gyromagnetic ratio to 9 decimal places, or so. The list of examples could go on for a long time, but let us use these four as a representative sample.

These triumphs of mathematical physics are important and very impressive, but are they "mysterious" or "miraculous"? The author thinks not. Is it not more reasonable to believe that our successful physical theories originate from the recognition of an apparent pattern that approximates the underlying order of nature, and that mathematics is used as a vehicle for formalizing and quantifying those patterns, or complex interrelationships. If the pattern is a good approximation to the actual underlying order of nature, and if the applied mathematics accurately embodies the pattern, then it should not be the least bit surprising that the pattern or its mathematical embodiment holds good beyond the empirical limits that are applicable at a given time and can thus be used to make successful predictions.

The development of electromagnetic theory is an excellent case in point. A "non-mathematical" (as he described himself) Faraday, through the natural philosophic techniques of observation, analogy and symmetry arguments, arrived at the basic qualitative ideas of the patterns involved in electromagnetic phenomena and invented the concept of a field ("lines of force") to elucidate a geometric and pictorial model of those patterns. Maxwell, in turn, discovered the mathematical formalism that would best capture Faraday's ideas, and showed some impressive creativity of his own in doing so. The patterns (or order) embodied in Maxwell's equations are good approximations to the actual laws of electromagnetic phenomena, and these patterns could be extended beyond existing observations to predict EM waves. In principle, at least the existence of these waves could have been predicted solely on the basis of Faraday's purely qualitative model. The "helium miracle", the Lamb shift and the accuracy of QED's prediction of g (see question 12 for a follow-up on this) can all be understood according to the same reasoning. We observe, we recognize patterns, we embody those patterns in mathematics, and we make predictions based on extensions of the patterns into the unknown. If all works well, the predictions are subsequently vindicated when the observational boundaries expand.

However, it is crucial to note that these inferred patterns and the mathematical "miracles" based upon them *always break down* at some level of refinement, or when the observational phenomena that spawned the pattern recognition become a sufficiently small subset of all relevant observations. This shows clearly that our patterns are not the actual patterns of nature's more sophisticated underlying order, and that the most we can "get out" of applied mathematics is an extension of a pattern that we have already built into the mathematics itself.

A very good case can be made for the contention that all forms of cognition are largely, or perhaps totally, due to various types of pattern recognition.[16] When one is able to investigate the original conceptual steps toward an important scientific discovery, direct or abstract pattern recognition usually figures very prominently in that early thinking; as the theory matures the initial intuitive attempts at pattern recognition are overshadowed by the mathematical formalism, which exudes elegance, rigor and a degree of intimidation.

(10.) Is Mathematics the Ultimate Form of Physical Knowledge?

Before attempting an answer to this question, I will reiterate my epistemological scenario for how physical knowledge comes into being. The basic steps in that epistemological scenario are: (1) careful and unbiased observation of natural phenomena, (2) *conceptual* recognition of overt or covert patterns or order that lead to a unifying conceptual paradigm for the relevant phenomena, and (3) the development of an analytical model that formalizes the conceptual paradigm and gives it scientific (quantitative) rigor. Often there is interplay among the steps and recycling of the 3-step process.

In physics the third step involves the introduction and refinement of a rigorous mathematical framework for the paradigm. But as Kuhn has successfully argued,[17] paradigms have limited applications and ironically it is the introduction of a mathematical formalism that both allows the paradigm to reach its full potential and yet sows the seeds for its eventual downfall by permitting a series of ever more "risky" predictions that will finally reveal the limits of the paradigm. A very important point is that when a paradigm fails and is superseded, it is not its mathematical formalism that usually plays the creative role in the development of the superior new paradigm. Rather, what is extracted from the old paradigm and used to discover the new paradigm is usually a small number of natural philosophic principles.[5] In the transition from Newtonian gravitation to general relativity, for example, Newton's mathematical formalism was exchanged for Riemann's, but the natural philosophy embodied in the concept of the universality of gravitation and the enigmatic equivalence of inertial and gravitational mass were carried over into the new paradigm and actually played a creative role in its development. The basic 3-step epistemological scenario for the development of a paradigm and the inevitable succession of increasingly more encompassing paradigms can be identified throughout the history of science, and from the scale of major scientific revolutions to the scale of very modest personal advances.

After the preliminary remarks given above, the author feels justified in arguing that the answer to question 10 is no. Applied mathematics may represent the most scientifically accurate form of the physical knowledge embodied in a *given* paradigm, but since the paradigms never last forever, it seems inappropriate to claim that mathematics is the ultimate form of physical knowledge. Rather it seems to me that the tools of mathematics and natural philosophy are equally important methods in an endless epistemological cycle that has no ultimate culmination (at best, it asymptotically approaches "truth"). Moreover, even in the case of a single paradigm the applied mathematics rests on a foundation of natural philosophy, and "initial conditions"[1] must be specified before the mathematics can be made to "work". The notion that physics, or the other sciences, can be completely reduced to an ultimate set of mathematical equations is a recurrent, idealistic delusion.

(11.) How Effective is Mathematics as a Tool for Discovery in the Natural Sciences?

According to the epistemological scenario given in the answer to question 10, mathematics plays a crucial role in elucidating and formalizing ideas in the natural sciences. But to what extent does mathematics play a leading role in directly stimulating new ideas and discoveries about nature? My position is that mathematics can be used as a discovery tool, and a couple of examples might be Dirac's work in quantum field theory[18] and Maxwell's mathematical elaboration of Faraday's EM paradigm. Mathematics also plays an important role in the discovery process by helping to reveal the limits of a paradigm through quantitative predictions. But it seems to me that the discovery process in the natural sciences is more typically aligned with the empirical and conceptual (pattern recognition) processes of natural philosophy. The development of modern EM theory provides a particularly clear-cut and compelling case in point,[19] and virtually all of Einstein's diverse contributions to physics follow the epistemological scenario presented above. The following quotation by Einstein is appropriate to the present discussion.

"The theorist's method involves his using as his foundation general posulates or "principles" from which he can deduce conclusions. His work thus falls into two parts. He must first discover his principles and then draw the conclusions which follow from them. For the second of these tasks he receives an admirable equipment at school. If, therefore, the first of his problems has already been solved for some field or for a complex of related phenomena, he is certain of success, provided his industry and intelligence are adequate. The first of these tasks, namely, that of establishing the principles which are to serve as the starting point of his deduction, is of an entirely different nature. Here there is no method capable of being learned and systematically applied so that it leads to the goal. The scientist has to work these general principles out of nature by perceiving in comprehensive complexes of empirical facts certain general features which permit precise formulation."[8]

When I look at the history of the natural sciences (especially physics and astronomy) and reflect on the scientific progress that has taken place in recent decades, I recognize the unquestionably important roles that mathematics has played, but it seems to me that mathematics usually follows rather than leads in the discovery of principles and patterns in natural phenomena.

(12.) Can Mathematics have Inhibitory Effects on the Development of Science?

While the benefits to science provided by mathematics are obvious and thoroughly chronicled, there are several ways in which applied mathematics can inhibit scientific progress, and these inhibitory aspects of applied mathematics are less well appreciated.

Firstly, there is the problem of choosing over-simplified mathematical models because they are more tractable. In my main area of interest, cosmology, this problem is quite serious. Almost two decades ago de Vaucouleurs made the following comment, and it is all too appropriate even today. "With few exceptions modern theories of cosmology have come to be variations on the homogeneous, isotropic models of general relativity. Other theories are usually referred to as "unorthodox", probably as a warning to students against heresy. When inhomogeneities are considered (if at all), they are treated as unimportant fluctuations amenable to first-order variational treatment. Mathematical complexity is an understandable justification, and economy or simplicity of hypotheses is a valid principle of scientific methodology; but submission of all assumptions to the test of empirical evidence is an even more compelling law of science."[20] The tendency to ignore the full character of natural phenomena because the complex patterns they suggest are difficult to formalize mathematically is a real threat to short-term progress in understanding nature. Eventually, one hopes, the phenomenological details that refuse to "go away" will be acknowledged and will enter, in one form or another, the body of scientific knowledge. The dogged determination to fit nature into over-simplified mathematical molds can be seen in the historical reluctance to move beyond Euclidean geometry and in the more recent difficulties in appreciating the fact that self-similar geometry is ubiquitous in nature.[21]

Secondly, when a flexible mathematical model is molded around preexisting observational data, one can often achieve stunning retrodictive feats that are too often mistaken for definitive predictions.[15] Given enough time and effort, one can "predict" the electron's gyromagnetic ratio to a remarkable number of significant digits by several *different* models.[22] The mathematical achievement of close quantitative agreement with preexisting observations can be quite beneficial to scientific progress, but it can also hinder that progress by fostering over-confidence. Definitive tests of a mathematical model can only come through unequivocal predictions.

Thirdly, if too much emphasis is placed on the mathematical aspects of modelling, then there can be a serious neglect of the natural philosophic assumptions that initially prescribed the choice of models.

Fourthly, there is the sheer intimidation of mathematics. A paper bristling with equations and mathematical jargon is almost invariably judged to have more competent ideas than a paper that lacks such esoterica. And yet the reverse might actually be the case.

(13.) Does Science Progress?

Given Proposition 3 that absolute knowledge is beyond our reach and the fact that theories and paradigms are eventually found wanting and are replaced, the notion that science progresses in the traditional sense has come under question. In his classic study of paradigmatic change in the sciences, *The Structure of Scientific Revolutions*,[18] Kuhn comments: "We may, to be more precise, have to relinquish the notion, explicit or implicit, that changes of paradigm carry scientists and those who learn from them any closer to the truth." While Kuhn argues that some forms of progress certainly occur and qualifies his doubts about the lack of "coherent direction of ontological development", others flatly reject the notion that nature has an objective reality (with objectively real entities and unique natural laws governing their behavior) and deny that science gradually achieves a better overall understanding of the objective reality of nature.

The author believes that science does progress in terms of accumulating more and better information about nature and in terms of "ontological development". Certainly mathematics, the language of science, has progressed a great deal over the last few centuries. Considering applied geometry, for example, the advent of non-Euclidean geometry represented a major step forward in the sophistication of geometry and the geometric modelling of nature. Likewise, the author suspects that the introduction of "fractal" or self-similar geometry[21] represents an equally important advance towards an understanding of the true complexity of nature's intrinsic geometry.

The author feels that there is also compelling evidence of progress in natural philosophy. Over the last few centuries there has been an enormous accumulation of empirical information about nature. Moreover, although paradigms and their applied mathematical formalisms tend to go the way of all things human, a very large number of ideas and principles of natural philosophy survive obsolescence and are carried forward to become the ever-deepening foundations of succeeding paradigms. A *few* of these ideas and principles are: the universality of interactions such as gravitation, the conservation of mass/energy, the "equivalence" of mass and energy, the atomic composition of matter, the principles of thermodynamics, the idea that the Sun is a star, the concept of biological evolution, etc. I regard this vast amount of knowledge as the natural philosophic legacy of past paradigms and contend that it represents approximate knowledge that will be refined, but not rescinded.

When I look at the last 400 years of scientific history it seems to me that individual theories and paradigms have become increasingly more able to reveal the underlying interconnection of apparently disparate natural phenomena and that our *overall* understanding of nature has become progressively more unified and coherent. I fail to comprehend how a philosopher of science can look at the history of science and not see real progress, unless one rejects the objective reality of nature. But in that case all of science would ultimately appear to be an arbitrary game, and something of a fool's game at that. It seems to me that those who would choose this latter path have conscious or subconscious motivations for belittling science, e.g., to raise their intellectual status or to decrease that of others.

(14.) Is it So Mysterious that the Universe is Highly Ordered?

Many authors who have discussed the math/science relationship have answered this question strongly in the affirmative. Wigner[1] stated: "it is not at all natural that "laws of nature" exist, much less that man is able to discover them." Hamming[2] comments: "Is it not remarkable that 6 sheep plus 7 sheep make 13 sheep;... Is it not a miracle that the universe is so constructed that such a simple abstraction as number is possible?"

While the author thinks that it is right and proper to feel a profound sense of awe at the infinite orderly complexity of the cosmos, perhaps a sense of bewilderment is much less useful. If one is going to argue that the answer to question 14 is yes, then one presupposes that a large number (infinite?) of different universes were possible, or currently exist. An alternative presupposition (Proposition 4) is that there is only one cosmos and that it could only be the way it is. In this case the answer to question 14 is no. Currently there is no *scientific* (i.e., testable in a definitive manner) way to decide which presupposition is correct (if either is), though some would argue (unjustifiably, I believe) that theoretical arguments make the former more likely. The author prefers the simplicity and clarity of the latter assumption.

As an aside, while it is true that the orderliness of the universe permits us to be here to observe it, to conclude that the universe was in some way specifically constrained to make that observation possible seems to me to be putting the cart before the horse.

References

1. E. P. Wigner, *The Unreasonable Effectiveness of Mathematics in the Natural Sciences, Commun. Pure Appl. Math.* **13** (1960) 1–14.
2. R. W. Hamming, *The Unreasonable Effectiveness of Mathematics, Am. Math. Monthly* **87** (1980) 81–90.
3. M. Kline, *Mathematics And the Search For Knowledge* (Oxford University Press, 1985).
4. G. H. Hardy, *A Mathematician's Apology* (Cambridge University Press, 1940).
5. J. Bronowski, *A Sense of the Future* (MIT Press, 1977).
6. H. Weyl, *Philosophy of Mathematics and Natural Science* (Atheneum, 1963).
7. J. Singh, *Great Ideas of Modern Mathematics* (Dover, 1959).
8. A. Einstein, *Ideas and Opinions* (Crown, 1954).
9. R. Courant and H. Robbins, *What is Mathematics?* (Oxford University Press, 1941).
10. J. Gleick, *Chaos* (Viking, 1987).
11. J. D. Barrow, *The Lore of Large Numbers: Some Historical Background to the Anthropic Principle, Quarterly J. Roy. Astron. Soc.* **22** (1981) pp. 388–420.
12. J. D. Barrow and F. J. Tipler, *The Anthropic Cosmological Principle* (Oxford University Press, 1983).
13. A. Einstein, *Mein Weltbilt* (Querido Verlag, 1934).
14. P. W. Bridgeman, *The Logic of Modern Physics* (Macmillan, 1946).
15. R. L. Oldershaw, *The New Physics – Physical or Mathematical Science?, Am. J. Phys.* **56** (1988) 1075–1081.
16. H. Margolis, *Patterns, Thinking, and Cognition* (University of Chicago Press, 1987).
17. T. S. Kuhn, *The Structure of Scientific Revolutions*, 2nd edition (University of Chicago Press, 1970).
18. A. Pais, *Inward Bound* (Oxford University Press, 1986).
19. R. L. Oldershaw, *Faraday, Maxwell, Einstein and Epistemology, Nature and System* **3** (1981) 99–108.
20. G. de Vaucouleurs, *The Case for a Hierarchical Cosmology, Science* **167** (1970) 1203–1213.
21. B. B. Mandelbrot, *The Fractal Geometry of Nature* (W. H. Freeman, 1982).
22. H. Aspden, *QED and Copernican Epicycles, Am. J. Phys.* **54** (1986) 1064.

Mathematics and the Language of Nature

F. David Peat
90 Fentiman Avenue, Ottawa,
Ontario KIS OT8, Canada

1. Introduction

"God is a mathematician", so said Sir James Jeans.[1] In a series of popular and influential books, written in the 1930s, the British astronomer and physicist suggested that the universe arises out of pure thought that is couched in the language of abstract mathematics. But why should God think only in mathematics? After all, some of most impressive achievements of the human race have involved architecture, poetry, drama and art. Could the essence of the universe not equally be captured in a symphony, or unfolded within a poem?

Three centuries earlier, Galileo had written, "Nature's great book is written in mathematical language" an opinion that has wholeheartedly been endorsed by physicists of our own time. Mathematics today occupies such an important position in physics that some commentators have argued that it has begun to lead and direct research in physics. In a frontier field, called Superstrings, some critics are arguing that mathematics is actually filling in the gaps left by the lack of any deep physical ideas. But why should mathematics play such a powerful role in physics? Is its central position inevitable? And is the present marriage between physics and mathematics always healthy, or are there ways in which mathematics may, at times, block creativity? In this essay I want to explore, in a speculative and free-wheeling way, some possible answers to these questions and to make some suggestions as to some radical developments in a language for the physical world.

2. The Role of Mathematics

While there are exceptions, it is generally true that great mathematics is studied for its own sake and without reference to anything outside itself. Mathematics has a beauty all its own and there is, for the mathematician, an aesthetic joy that comes from solving an important problem, no matter what value society may place on this activity. In this sense, mathematics has constantly sought to free itself from its practical origins.

Geometry, for example, began with rules for surveying, calculating the areas of fields and making astronomical studies and acts of navigation. Probability theory had its origins in the desire to raise gambling to a high art. But, very quickly, mathematics shook itself free from such pedestrian origins. While it is certainly true that some exceptional mathematicians have begun their studies with a concrete problem taken from the physical world, in the end, the mathematics they have developed has moved away from these specific cases in order to focus on more abstract relationships. Mathematics is not really concerned with specific cases but with the abstract relationships of thought that spring from these particular instances. Indeed, mathematics takes a further step of abstraction by investigating the relations between these relationships. In this fashion, the whole field moves away from its historical origins, towards greater abstraction and increasing beauty.

The English mathematician G. Hardy[2] refused to justify mathematics in terms of its utility and pursued it as an art for its own sake. He seemed to rejoice in the very abstraction of his own research and in its remoteness from practical applications. Indeed, Hardy once spoke of a monument so high that no one would ever be able to see the statue that was placed at its pinnacle – a fitting metaphor for his own, somewhat extreme, view of the role of mathematics.

In von Neumann words, mathematics is "the relation of relationships". Today it is possible to go further, for that branch of mathematics called Category theory is not concerned with any particular field of mathematics but with the relationships between the different fields themselves! Mathematics at this level has the appearance of the purest and most rarefied thought. It is like a piece of music of such abstract perfection that the realization of a single performance would destroy its purity.

But it is exactly at this point that a staggering paradox hits us in the teeth. For abstract mathematics happens to work. It is useful. It is the tool that physicists employ in working with the nuts and bolts of the universe! Indeed, scientists of the old school referred to mathematics as "the handmaid of physics". But why should an abstract codification of pure thought, divorced from any reference to physical objects and material processes, be so useful in the daily practice of science? To echo Eugene Wigner's famous remark, mathematics is *unreasonably effective*.

There are many examples, from the history of science, of a branch of pure mathematics which, decades after its invention, suddenly finds a use in physics. There are also cases of a mathematical approach, developed for one specific purpose, that is later found to be exactly what is needed for some totally different area of physics.

Probability theory, first devised to deal with strategies of gambling, ends up as the exact language needed to give a molecular foundation to thermodynamics – the physical theory dealing with work and heat. But why should this be so? When Einstein formulated his general theory of relativity he discovered that the necessary mathematics had already been developed in the previous century. Similarly the mathematics required for quantum theory was ready and waiting. Group theory, the cornerstone of much of theoretical physical of the last fifty years, had its origins in fundamental mathematics of the 18th and 19th centuries. And, when it comes to Superstrings, a topic at the frontiers of contemporary theoretical physics, the mathematical tools of cohomology and differential geometry are waiting to be used. On the face of its, this apparently perfect marriage between abstract mathematics and the study of the physical world is as improbable as discovering that a piece of modern sculpture fits exactly as the missing component of some complex new engine!

How is it possible to account for this unreasonable effectiveness of mathematics and for the powerful role it plays in physics today? One approach is to take the hint offered by Galileo and view mathematics as a language. Just as natural language is used for everyday thought and communication, so too, physics has to make use of whatever mathematical languages happens to be lying around. Mathematics, in this view, is a tool and, like the hammer or screwdriver, we select the available tool that best fits the job.

3. Mathematics as Language

It is common to talk of "the language of mathematics". But is mathematics really a language? Does it possess the various properties that are characteristic of other natural languages? Clearly mathematics does not have the same fluency as a natural language and, even more obviously, it is rarely spoken aloud. This suggests that mathematics is really a more restrictive limited form of language. Nevertheless, the suggestion is that everything mathematics can do must ultimately find its origin in language. This means that the rich and abstract proofs and theorems of mathematics can ultimately be traced back to thoughts and arguments that were once voiced in language – albeit in a long winded and cumbersome way. Now, it is obvious that mathematics does not look anything like natural language. Mathematics deals with numbers and symbols, it is used to make calculations and its form is highly abstract. On the other hand, all these features may already be enfolded within natural language. The power of language lies in the way meaning can be conveyed through form and transformation. The Ancient Greeks, for example, realized that truth could be arrived at through various patterns of sentences.

All men are mortal.

Socrates is a man.

Therefore, *Socrates is mortal.*

Or, to take another pattern,

Some mathematicians are clever.

All mathematicians are animals.

Therefore, *Some animals are clever.*

What is striking about these patterns is that the truth of the conclusion does not depend on the content of the sentences but on their form. In other words, substitutions do not affect the validity of the proof

All [*cats*] are [*wanderers*].

[*Minou*] is a [*cat*].

Therefore, [*Minou*] is a [*wanderer*].

Clearly these patterns and substitutions have something in common with algebra. Other transformations are also possible within language.

From

John shut the door

we get

The door was shut by John

These are only a few if the great range of abstract operations possible within language. Indeed, the linguist Noam Chomsky[3] has argued that this ability arises genetically and is inherent in all human thought. To take Chomsky's idea even further, we could say that mathematics has isolated and refined several of the abstract elements that are essential to all human languages. An extreme form of this argument would be to say that while mathematicians may make abstract discoveries and develop new mathematical forms, in the last analysis they simply represent something that is inherent in human thought and language.

The normal way we express and communicate our thoughts is through language and mathematics becomes a formal extension of this process. So when physicists seek a rational language in which to express their insights, they simply take what happens to be at hand – the best available mathematics. It is not therefore surprising that mathematics happens to work.

Mathematics has played a vital role in raising the speculations an earlier age to the highest peaks of intellectual enquiry. But I am now putting forward the hypothesis that physicists have, in fact, no alternative. Mathematics has been forced on them as the only language of communication which can also serve to make, with precision and economy, quantitative predictions and comparisons. And, when no Issac Newton happens to be around to develop a new mathematical language hand in hand with new physical insights, then physics has to make to do with what is available.

In those cases in which the form of the mathematical language makes a perfect marriage with the content of the physical ideas, the communication and developement of physics is highly successful. But this may not always be the case. Sometimes it may turn out that a particular mathematical language is forced, by physics, to say things in cumbersome ways. The mathematics actually gets in the way of further creativity. At the other extreme, it is the very ease of expression that drives a theory in a particular direction so that mathematics actually directs the evolution of physics, even when new physical insights are lacking. In other words, I want to question Wigner's claim that mathematics is *unreasonably effective*. For it could be that the whole thing is an illusion brought about because physics has no other language in which to communicate quantitative statements about the world. In the past decades, there has been much talk about paradigm shifts and scientific revolutions – yet it is still possible to retain the same mathematical language after such a radical shift. In short, the whole baggage of unexamined presuppositions that are inherent in the mathematics are carried over to the new physics.

Any writer knows that language has the power to take over his or her ideas. Words have their own magic, and a style, once adopted, will gather its own momentum. It has been said that a writer is possessed by all the texts that have been previously written. As soon as we put pen to paper and chose a particular literary form then what we write is, to some extent, predetermined. I would suggest that the same is true of physics. That the adoption of a particular mathematical language will subtly direct the development of new ideas. Moreover, there are times when mathematics may actually block the operation of a free, creative imagination in physics. Since mathematics occupies such a prominent place in physics today, these are vital questions to be explored.

In arguing that mathematical languages direct and influence our thought in science, we now see that the real danger arises from always focusing on the physical ideas and not giving attention to the language in which they are expressed! As long as physicists view mathematics simply as a tool then it is possible to ignore the subtle but very powerful influence it has over the way they think and how they express their thoughts. In fact, I believe that a good argument can be made that a particular form of mathematics has been blocking progress in physics for decades – this is the Cartesian co-ordinate system, a mathematical form that has survived several scientific revolutions!

A major problem facing modern physics is that of unifying quantum theory with relativity. One theory deals with discrete, quantized processes below the level of the atom. The other with the properties of a continuous spacetime. While it is certainly true that deep physical issues must be resolved before significant progress can be made, I would also argue that the mathematical language in which the quantum theory is expressed is at odds with what the theory is actually saying. While quantum mechanics and quantum field theory are truly revolutionary approaches, the mathematics they are based on goes right back to Descartes – to the same Cartesian co-ordinates we all learned at school. For three hundred years, physics has employed the language of co-ordinates to discuss the movement of objects in space and time. Later developments like the calculus also rely upon this idea that space can be represented by a grid of co-ordinates. But it is this same mathematical language that is at odds with the revolutionary insights of quantum theory. Cartesian co-ordinates imply continuity, as well as the notion of space as a backdrop against which objects move. So whatever new insights physics may have in this area, they are still being expressed in an inappropriate language. This, I believe, represents a major block to thinking about space and quantum processes in radically new ways.

The example of how the Cartesian grid has dominated physics is rather obvious. But there may be many other, and more subtle, ways in which particular mathematical forms are currently directing science and limiting the possibilities for its development.

4. Mathematics beyond Language

But is it really true that mathematics is nothing more than a limited and abstract version of natural language? I would argue that mathematics is both more, and less, than a language. Since it involves highly codified forms, mathematics makes it easy to carry out calculations, to demonstrate proofs and to arrive at true assertions. But, in my opinion, this is only a surface difference, a feature of the convenience and economy of mathematics over ordinary language. A more significant way in which mathematics goes beyond language is that it involves a particular kind of visual and sensory motor thinking that does not seem to be characteristic of ordinary language. Some parts of mathematics deal with the properties and relationships of shapes. While these properties can be generalized to many dimensions and to highly abstract relationships, nevertheless, mathematicians have told me that their thinking in these particular fields enter regions which do not involve language in any way. It calls upon a sort of direct, internal visualization and may even involve an internal sense of movement and of tiny muscular reactions. This "non-verbal" thinking may also take place in other fields of mathematics and appears to involve a form of mental activity that goes beyond anything in the domain of a spoken or written language. It could be that, at such times, mathematical thought has direct access to a form of thinking that is deeper and more primitive than anything available in any natural language. This pre-linguistic mental activity may be the common source from which both mathematics and ordinary language emerge.

On the other hand, mathematics is also less than a language, in that it lacks the richness, the ability to deal with nuance, the inherent ambiguity and the rich strategies for dealing with this ambiguity. In this sense, mathematics is a limited, technical language in which much that is of deep human value cannot be expressed.

5. Mathematics and Music

It is possible to explore the nature of mathematics, and its relationship to physics, in another direction by comparing it to music. Mathematics is an abstract system of ordered and structured thought, existing for its own sake. It is possible to apply a similar description to music. Indeed the 20th century composer, Edgar Varese, has written that "music is the corporealization of thought". Listening to Bach, for example, is to experience directly the ordered unfolding of a great mind. This suggests that music and mathematics could be related in some essential way. On the other hand, who would employ music to express a new theory of the universe? (But could this simply be a prejudice that is characteristic of our earth-bound consciousness? Do beings in some remote corner of the universe explore the nature of the universe in music and art?)

Music and mathematics are similar, yet different. Indeed, I believe that both the strengths and the weakness of mathematics lie in this difference. Mathematics has developed to deal with proof and logical truth in a precise and economical way. Mathematics also makes a direct correspondence with the physical world through number, calculations and quantitative predictions.

While it could be said that music is "true" in some poetical sense and that the development of a fugue has a logical ordering that is similar to that of a mathematical proof, on the other hand, these are not the primary goals of music. Music deals with the orders of rational thought, yet it is also concerned with the exploration of tension and resolution, with anticipation, with the control of complex sensations of sound and with the evolution and contrast of orders emotion and feeling. To borrow a Jungian term, music could be said to be more complete, for it seeks a harmony between the four basic *human functions*; thought balanced by feeling and intuition by sensation. While mathematicians may experience deep emotions when working on a fundamental piece of mathematics, unlike composers, their study, *per se*, is not really concerned with the rational ordering of these emotions or with the relationships between them. The greatest music, however, moves us in a deep way and leaves us feeling whole. It engages thought and emotion, it expresses itself through the physical sensation of sound.

In this sense it could be said that physics, with its reliance on the language of mathematics, must always present an incomplete picture of the universe. Its language is impoverished, for it lacks this basic integration of the four human functions. It can never fully express the essential fact of our confrontation with, participation in, and understanding of nature.

But is it possible, I wonder, that, in the distant future, science, inspired by the example of music, may develop a more integrated and versatile language, one which would have room, perhaps, for the order of emotion and direct sensation while, at the same time, retaining all the power of a more conventional mathematics?

There is yet another significant way in which "the language of music", and of the other arts, differs from mathematics. While all these languages are concerned with relationships and rational orders of thought, the arts are able to unfold these orders in a more dynamical way by exploring the way order is generated in the act of perception itself. Quantum theory is also concerned with the indivisible link between the observer and the observed. And this suggests that it would be to the advantage of physics to develop a similar flexibility in its basic language giving it the ability to explore the rich orders that lie between the observer and the observed.

Let me explain what I mean. A great work of art possesses a rich internal order. In music, for example, a theme may be transposed, inverted, played backwards and otherwise transformed in a variety of ways which still retain a certain element of its order. Of course, this is only one simple example of the sorts of order explored in a musical composition, indeed the order of great music is so rich as to defy complete analysis. Likewise, a painting contains complex relationships between its lines, masses, areas, colors, movements and so on. In some cases such objective orders may have much in common with the sorts of order that are found in mathematics. But what makes any work of art come alive is its contemplation by the human observer:- Music played in a vacuum is not music, art that is never seen is not art. For the work of art arises in that dynamic interaction between the active perception, intelligence, knowledge and feeling of the viewer and the work itself.

To take a particular example, some of the drawings of an artist like Rembrant, Picasso, Mattise or a Japanese master appear, on the surface, to be extraordinarily simple. Few marks appear on the paper when contrasted with, for example, the detailed rendering done by an art student. A trivial analysis would suggest that the sketch contains "less information" than the detailed rendering and that its order is relatively impoverished. Yet the confrontation of a viewer with a Mattise drawing is a far richer experience in which complex orders of thought and perception are evoked. To make the slightest change in position, direction, gesture or even thickness of a single line can destroy the balance and value of a great drawing, but may have only a negligible effect on a student work. In this sense, great art has an order of such richness, subtlety and complexity that it is beyond anything that can be addressed in current mathematics. Yet it is something to which the trained viewer can immediately respond.

Indeed, the rich order of the drawing lies not so much in some objective order of the surface marks on the paper, but in the whole act of perception itself and in the way in which the drawing generates a hierarchy of orders within the mind. Lines evoke anticipations in the mind that may be fulfilled in harmonious or in unexpected ways. The mind is constantly filling in, completing, creating endless complex orders. A single line may suggest the boundary of a shadow, the outline of a back or it may complete a rhythm created by other lines. Indeed the act of viewing a drawing could be said to evoke an echo, or resonance, of the whole generative process by which the drawing itself was originally made. The essence of the drawing does not therefore lie in a static, objective order – the sort of thing that can be the subject of a crude computer analysis involving the position and direction of a number of lines. Rather, it is a rich dynamical order, an order of generation within the mind. Through his or her art, the creator of the drawing has called upon the nature of the subject, the history of art, and on all the strategies that are employed in perception. So standing before a drawing involves a deep and complex interplay between the work itself, the visual center of the brain, memory, experience, knowledge of other paintings, and of the human form. The eyes, memory, mind and even the body's sensory-motor system become involved in the generation of a highly complex order, and order in which every nuance of the drawing has its part.

The order within an economical drawing may, therefore, be far richer than we first suspect. For its power lies not so much in some surface pattern of the lines but in the controlled and predetermined way in which these lines generate, through the act of perception itself, infinite orders within the mind and body. While attention has certainly been given, by researchers in Artificial Intelligence, to what is called *the early processes of vision*, it is clear that the sort of order I am talking about lies far beyond anything that mathematics or aritificial intelligence could analyze or even attempt to deal with at present.

I feel that the description of complex orders of perception and generation is a rich and powerful area into which mathematics should expand. It may also have an important role to play in physics. Quantum theory, for example, is concerned with the indissoluble link between observer and observed and it would be interesting to make use of a mathematics which can express the infinite orders that are inherent in this notion of wholeness.

A similar sort of argument applies to music. Some musicologists have gone so far as to analyze music by computer, and to calculate its "information content", concluding, for example, that "modern music" contains more information than baroque music! But the essence of music does not lie in some measure of its objective information content but in the rich and subtle activity it evokes within the mind. Music and art are seeds that, in a controlled and deliberate way, generate a flowering of order and meaning within the mind and body of the listener.

To return to an earlier point; this generative order suggests a reason why great music could indeed act as a metaphor for a theory of the universe. Music is concerned with the creation and ordering of a cosmos of thought, feeling, intuition and sensation and with the infinite dynamical orders that are present within this cosmos. In this sense, music could be said to echo the generation and evolution of a universe. Clearly our present mathematics lacks this essential dimension. But could mathematics, in fact, move in such a direction? A new mathematics would not simply offer a crystallization of thought but also explore the actual generative activity of the orders of this thought within the body and mind. Such a new formal language would represent a deep marriage between mathematics and the arts. It would involve a mathematics that requires the existence of another mind to complete it, in an ordered and controlled way, and, in so doing, this mathematics would become the germ of some, much deeper order.

6. Mathematics and the Brain

Let us return again to the question of the unreasonable effectiveness of mathematics. As we have seen, one answer is to consider mathematics as a language, indeed the only available language that can deal, in an economical and precise way, with quantitative deductions about the world. Mathematics, in this sense, is a restricted form of natural language. But, in other ways, it goes beyond language. Physics, however, is always in the position of being forced to use mathematics to communicate at the formal level. The question, therefore, is not so much one of the unreasonable effectiveness of mathematics, but of physicists having no real alternatives.

But there may be other ways of looking at this question. One way is to suggest that mathematics, in its orders and relationships, is a reflection of the internal structure and processes of the brain. In moving towards the foundations of mathematics one would therefore be approaching some sort of direct expression of the controlling activities of the brain itself. And, since the brain is a physical organ that has evolved through its interactions with the material world, it is inevitable that the brain's underlying processes should model that world in a relatively successful way. Human consciousness has developed, in part, as an expression of our particular size and scale within the environment of our planet. It is a function of the particular ranges of senses our bodies employ, and of our need to anticipate, plan ahead, hold onto the image of a goal and remember. Moreover, consciousness has created, and been formed by, society and the need to communicate. It has brought us to the point where we can ask, for example, if we think because we have language or, if we have language because we think? Or if the answer could lie somewhere in between.

According to this general argument, the brain's function is a direct consequence of, and a reflection of, our particular status as physical and social beings on this planet. Mathematics, moreover, is a symbolic expression of certain of the ordered operations of this brain. It should come as no surprise, therefore, that mathematics should serve as a suitable language in which to express the theoretical models that have been created by this same brain.

This whole question of the formal strategies employed by the brain is the province of cognitive psychology. One of the pioneers in that field was Jean Piaget.[5] Piaget's particular approach was to suggest that the basis of our thought and action could be traced to the logic of the various physical transactions we had with the world during our first weeks, months and years. Piaget believed that these same logical operations are also present in mathematics and, in this respect, he had a very interesting point to make. It is well known, he pointed out, that mathematics can be arranged in a hierarchical structure of greater and greater depth. In the case of geometry, for example, the top, and most superficial, level is occupied by those semi-empirical rules for surveying and calculating shapes that were known to the Egyptians and Babylonians. Below that could be placed the more fundamental, axiomatic methods of the ancient Greeks. The history of geometry demonstrates the discovery of deeper and more general levels, Euclidean geometry gives way to non-Euclidean, beneath geometry is topology, and topology itself is founded on even more general and beautiful mathematics. The longer a particular topic has been studied, the deeper mathematicians are able to move towards its foundations.

But Piaget, pointed out, this historical evolution is a direct reversal of the actual development of concepts of space in the infant. To the young child, the distinction between intersecting and non-intersecting figures is more immediate than between, say, a triangle, square and circle. To the infant's developing mind, topology comes before geometry. In general, deeper and more fundamental logical operations are developed earlier than more specific rules and applications. The history of mathematics, which is generally taken as a process of moving towards deeper and more general levels of thought, could also be thought of as a process of excavation which attempts to uncover the earliest operations of thought in infancy. According to this argument, the very first operations exist at a pre-conscious level so that the more fundamental a logical operation happens to be, the earlier it was developed by the infant and the deeper it has become buried in the mind. Again, this suggests a reason why mathematics is so unreasonably effective, for the deeper it goes the more it becomes a formal expression of the ways in which we interact with, and learn about, the world.

But, it could be objected, if the history of mathematics and, to some extent of theoretical physics, is simply that of uncovering, and formalizing, what we already know then how is it possible to create new ideas, like Einstein's relativity, that totally lie outside our experience? The point is, however, that this equality or interdependence of space and time was already present in all the world's language. Rather than coming to the revelation that time and space must be unified they have never really been linguistically separated! According to this general idea, what may appear to be novel in physics and mathematics is essentially the explicit unfolding of something that is already implicit within the structuring of human thought – of course physics itself also makes use of empirical observations and predictions. For this reason, the intelligent use of mathematics as a language for physics will necessarily make sense.

Piaget's notion, that the evolution of mathematics and physics is forever reaching the deepest structures of the mind, is certainly interesting. However, I feel that there is a certain limitation in the approach of cognitive psychology, with its emphasis upon strategies and programs of the brain, on successions of logical steps and on algorithms of thought. There is not sufficient space in this article to develop any detailed arguments, but I believe that, while cognitive psychology may produce some valuable insights, in its present form it does not capture the true nature of human intelligence in general, and mathematics in particular. Formal logic is an impoverished way of describing human thought and the practice of mathematics goes far beyond a set of algorithmic rules. The mathematician Roger Penrose[7] has, for example, produced compelling arguments why machine intelligence must be limited – a Turning machine, or indeed any other algorithmic device, will never be able to carry out all the sorts of things that a human mathematician can do. Mathematics may indeed reflect the operations of the brain, but both brain and mind are far richer in their nature than is suggested by any structure of algorithms and logical operations.

7. Mathematics and Archetypes

In this *final* section I am going to become more speculative and explore yet another approach to the question of the unreasonable effectiveness of mathematics. I want to suggest that mind and matter, brain and consciousness are two sides of a single process, something that emerges out of a deeper and hitherto unexplored ground. In this sense, the order of generation that gives rise to the universe has a common source with the generative order of consciousness. In its deepest operation, therefore, our intelligence could be said to mirror the world. But what can one say about the nature of this source? According to the classical Chinese philosopher, Lau Tzu, "the Tao which has a name in not the Tao", which seems to say it all.

Of course, the idea of an unknown, unconditioned source which is the origin of matter and consciousness may seem far fetched to many readers. But it is, after all, simply another way of accounting for the unreasonable effectiveness of mathematics. Our own age is out of sympathy with such sweeping assertions as "God is a mathematician", but suppose one suggests that mind and the universe have a common order and that the source of material and mental existence lies in a sort of unconditioned creativity, and in the generation of orders of infinite subtlety and complexity[8]? While the nature of such an order may never be explicitly known in its entirety, it may still be possible to unfold certain of its aspects through music, art and mathematics. The great aesthetic joy of mathematics is not, therefore, far from the joy of music or any great art, for it arises in that sense of contact with something much greater than ourselves, with the heart of the universe itself. Mathematics is effective when it becomes a hymn to this underlying order of consciousness and the universe, and when it expresses something of the truth inherent in nature.

This idea has been expressed in other ways. Carl Jung, for example, spoke of the archetypes. This is a difficult concept to convey in a short definition but, very roughly, the archetypes could be taken as those dynamical orders, unknowable in themselves, that underlie the structure of the collective unconscious. The archetypes are never seen directly but their power can be experienced in certain universal symbols. In his more speculative moments, Jung also hinted at something that lay beyond matter and mind, but included both. This psychoid, as he called it, is related to the archetypes and suggests that the same underlying ordering principles give birth and structure to both matter and mind. Just as human consciousness arises out of the collective unconscious, so too the universe itself arises out of something more primitive. Again we meet this notion that the same underlying order gives rise to both matter and mind.

Of particular interest is the importance that Jung placed upon numbers. Numbers, according to Jung, are direct manifestations of the archetypes and must therefore be echos of the basic structuring processes of the universe itself. It is certainly true that numbers are mysterious things. To return, for a moment, to the connection between mathematics and language. When it comes to language, it is a basic axiom of linguistics that "that sign is arbitrary". In other words, the meaning of a world does not lie in how it sounds or the way it is written but in the way it is used. If you want to know the meaning, the philosopher Wittgenstein said, look for the use. By contrast, the basic units of mathematics, the numbers, are totally different, they are not arbitrary but have a meaning and existence of their own. While the names given to the numbers may be arbitrary, the numbers themselves are not, 0, 1, 2, 3, are not symbols whose meaning changes with time and use but are the givens of mathematics. In a sense they are almost Platonic. It has been said, for example, that God made the numbers and the rest of mathematics is the creation of human intelligence. It is these same numbers that, Jung claims, are manifestations of the archetypes. Indeed Jung's argument does have a ring of truth about it for numbers are certainly curious things and the unfolding of their properties remains one of the most basic forms of mathematics. Could it be true, as the Jungians suggest, that the numbers are expressions of the archetypes or orders that underlie the universe and human consciousness[8]?

Curiously enough, this idea may have found favor with one famous mathematician. One of the most brilliant pure mathematicians in this century, S. Ramanujan, gave little value to mathematical proof but appeared to arrive at his remarkable theorems in number theory by pure intuition alone. Ramanujan himself, believed that these profound results were given to him by a female deity. In Jung's terminology, this deity would also be a manifestation of the archetypes.

So, to Ramanujan, the whole order of mathematics, with its underlying truth and beauty, essentially lies in a domain beyond logical truth and rational argument. It is something which can, at times, be touched directly by the mathematician's intuition and in a way that appears almost sacred. As to the nature of this domain, we can call it the archetypes, psychoid, ground of being or unconditioned, creative source. But what does it matter? What counts is that a remarkable mathematician bypassed rational argument and the need for vigorous proof and picked out outstanding theorems out of the air. And what is equally staggering is that, in all likelihood, these symphonies of pure thought may one day have totally practical applications in the real word.

8. Conclusion

The unreasonable effectiveness of mathematics remains an open question, although I have given some suggestions as to why it appears to work. I have also argued that mathematics may not always be as effective as we suppose, for physical ideas are sometimes forced to fit a particular mathematical language, in other cases the very facility of the language itself may drive physics forward, irrespective of any new physical ideas!

I have also suggested ways in which improvements in the formal language of physics could be advanced. A major area would be to discover a mathematics of complex and subtle orders, a formal way of describing what seems, to me, to be an essential feature of the universe. In addition these orders should also include the generation of order in the act of observation itself. There have recently been several attempts to describe complex orders – Mandelbrot's fractal theory is capable of describing and generating figures of infinite complexity; David Bohm's notion of the implicate order is a powerful concept but has yet to find an appropriate mathematical expression.[9]

Finally, I have also argued that there are times when the mathematical language of physics fails to capture the essential fact of our being in the universe. And here I must reveal another prejudice. Physics, to me, has always been concerned with understanding the nature of the universe we live in; a way of celebrating and coming to terms with our existence in the material world, rather than a matter of discovering new technologies and accumulating more knowledge. It is in this light that I have criticized the role of mathematics in physics and have hinted at the way new language forms could be developed. Of course I acknowledge the great service that mathematics has done for physics, how it has lifted it from speculation to precision, and, of course, I recognize the great power and beauty of mathematics that is practiced for its own sake. But here, at the end of the 20th century we must not rest on our laurels, the whole aim of our enterprise is to penetrate ever deeper, to move towards a more fundamental understanding and a more complete celebration of the universe itself. In this undertaking in which prediction, calculation and control over the physical world also have a place but they do not become the whole goal of the scientific enterprise. It is for this reason that I am urging physicists to pay closer attention to the mathematical language they use every day.

This whole concern with discovering and portraying the complex orders of nature, was also a preoccupation of the writer Virginia Woolf. Virginia Woolf was concerned with the order of the moment, with crystallizing, in language, the complex sensations, experiences and memories that make up each instant in a person's life. She recognized that, in the last analysis, the success of this enterprise depends on creating a fitting means of expression, on language, on words. Her own observations on this process convey precisely what I have been attempting to say in this essay:

"Life is not a series of gig-lamps symmetrically arranged; but a luminous halo, a semi-transparent envelope surrounding us from the beginning of consciousness to the end. Is it not the task to the novelist to convey this varying, this unknown and uncircumscribed spirit, whatever aberration or complexity it may display, with as little mixture of the alien as possible?"

For James Joyce it is the epiphanies or transcendent moments of life that have a special richness. They can occur at any instant and it is the business of language to capture these, even "transmuting the daily bread of experience into the radiant body of evolving life". For Virginia Woolf this radiant force of the moment must be captured by language "it is or will become a revelation of some order; is a token of some real thing behind appearances; and I make it real by putting it into words".

References

1. See, for example, J. Jeans, *The Mysterious Universe* (Cambridge University Press, 1930).
2. G. H. Hardy, *A Mathematician's Apology* (Cambridge University Press, 1967).
3. N. Chomsky, *Syntatic Structures* (Mouton Pubs, 1957).
4. For a preliminary discussion on the role of language in science see, A. J. Ford and F. D. Peat, *Foundations of Physics* (1988) 1233–1242.
5. See, for example, J. Piaget, *Structuralism* (Harper & Row, 1971).
6. M. L. von Franz, *Number and Time* (Northwestern University Press, 1974).

7. R. Penrose, *Question Physics and Conscious Thought*, in *Quantum Implications: Essays in Honour of David Bohm*, eds. B. J. Hiley and F. David Peat (Routledge and KeganPaul, 1987). See also, Penrose, *The Emperor's New Mind* (Oxford University Press, 1989).
8. See, for example, F. David Peat, *Synchronicity: the Bridge between Matter and Mind* (Bantam, 1987).
9. A discussion of complex orders is given in, D. Bohm and F. D. Peat, *Science, Order and Creativity* (Bantam, 1987).

The Reason Within and the Reason Without

John Polkinghorne
Queens' College, Cambridge
CB3 9ET, England
United Kingdom

The properties of subatomic matter are currently understood in terms of what is called the "standard model".[1] This is obtained by the juxtaposition of two theories, each successful within its own limited domain. One is quantum chromodynamics (QCD), the theory of the interactions of those quarks and gluons which are believed to be the constituents of nuclear matter. The second theory is the unified theory of Salam and Weinberg, which synthesized weak and electromagnetic interactions into a single scheme. This latter applies to a wider range of constituents, for it also brings within its account particles like electrons, which have no strong nuclear interactions, as well as the carriers of weak and electromagnetic interactions, the photon and the W and Z.

The attainment of the standard model was a very substantial intellectual achievement, resulting from experimental investigations and theoretical speculations stretching over a period of more than twenty years. Throughout, the theorists used symmetry as one of their principal guides – the notion that underlying the often apparently chaotic appearance of nature would be found a pattern which achieved elegance by its recognition that certain basic entities were to be treated in a similar, or symmetrical, fashion. Penetration to this basic pattern required imaginative leaps, for an essential ingredient in much theoretical thinking proved to be the idea of spontaneous symmetry breaking – the notion that the particles actually perceived might display a lesser degree of symmetrical perfection than the structure that underlay them. Mathematics provides the natural language by means of which to speak of symmetry. We can express the general concept by requiring that nothing essential is actually changed under the effect of certain transformations. (For example the symmetry of the circle is expressed by the fact that it remains unaltered in form by a rotation through any angle; symmetry between particles by the fact that their interchange produces no change in what is happening.) The branch of mathematics which studies such transformations is group theory and one could write a history of theoretical physics during the period in which the standard model was coming into existence in terms of the developing engagement of physicists with the theory of groups in the widest sense. When an account of internal transformations (for instance, linking one kind of quark with another) is embedded in the most flexible way possible within the setting of the spacetime continuum, one obtains a structure which is called a gauge field theory. The theories of modern elementary particle physics are all gauge field theories.

Quantum chromodynamics is such a theory, deriving its name from the fact that it refers to the so-called 'color' degrees of freedom of the quarks. The associated group is a mathematically attractive entity (technically, it is the semi-simple Lie group SU(3)). The Salam-Weinberg theory is not quite so economic or elegant for it involves two groups (technically, the direct product U(1)×SU(2)). The combined theory of the standard model is therefore less economic again, since it involves three groups simply juxtaposed in a direct product. A consequence of this loose form of association is that there are 19 adjustable parameters which have to be "put in by hand" before an attempt can be made to correlate the predictions of the model with experiment.

The standard model is phenomenologically successful, but no one feels quite content with it as a candidate for an ultimate theory of the structure of matter. There is a two-fold reason for this discontent. One point is that the theory does not bring within its embrace the fourth basic force observed in nature, namely gravity. For the phenomena of elementary particle physics investigated in the laboratory that force is far too weak to be of direct significance, but surely there must be a unified account of physical reality which successfully incorporates it with the rest. Such a combination has proved very difficult to attain. Even the synthesis of general relativity (the modern theory of gravity due to Einstein and itself based on a powerful use of the idea of gauge symmetry) and quantum theory has proved extremely difficult to formulate in a consistent fashion. The current agitation about theories of superstrings is due to the hope that they will enable this shortcoming to be remedied.

The other point of discontent is the mathematical ugliness of the standard model – all those groups and all those parameters. We have learnt to expect a greater economy and elegance in truly fundamental physical theory. That intuition has encouraged the search for Grand Unified Theories (GUTs for short) in which the standard model is embedded within a single larger group,whose structure then determines as many as possible of the otherwise adjustable parameters.

It cannot be said that either GUTs or superstrings have yet scored any unequivocal successes. This is not only in terms of phenomenological consequences but also in terms of formal persuasiveness. No scheme has yet been found having about it the majestic mathematical beauty which would encourage the thought that "surely this must be right". Yet the search continues, undeterred by the acknowledged fact that the regimes in which either a GUT or a superstring theory would manifest their true natures lie many orders of magnitude beyond what it is experimentally conceivable to investigate in the terrestrial laboratory. The only guides available to the huge army of talented physicists engaged upon this search are certain speculative interpretations of the hypothesised very-early history of the universe (at cosmic ages of much less than 10^{-30} sec) and – above all – the principle of seeking a mathematically beautiful expression for the theory. On this basis, theorists are content to devote themselves to an immense programme of exploratory work.

This story powerfully illustrates the faith that physicists have in what Eugene Wigner, in his celebrated Courant Lecture,[2] called "the unreasonable effectiveness of mathematics". Like most commitments of faith, it is by no means rationally unmotivated, even if it goes beyond what is rationally demonstrable.

Writing about the early days of modern quantum theory, Paul Dirac said about himself and Schrödinger that

> It was a sort of act of faith with us that any equations which describe fundamental laws of Nature must have great mathematical beauty in them. It was a very profitable religion to hold and can be considered as the basis of much of our success.[3]

The commitment to mathematical beauty as a guiding principle in theoretical physics was central for Dirac. He once expressed the opinion that it was more important to have this quality present in one's equations than that their solutions should fit with experiment. By this he did not mean to deny the importance of eventual empirical adequacy, but he wished simply to observe that its absence at some stage might merely indicate an inadequate approximation or an inaccurate experiment. On the other hand, mathematical ugliness would be a feature beyond redemption.

That feeling is widespread. A theoretical physicist faced with an inelegant and contrived theory will instinctively feel that it cannot be right. Equally, a compact and attractive proposal has about it an air of almost self-authentication. When one considers Maxwell's equations for electromagnetism, or Dirac's equation for the electron, or Einstein's equations of general relativity, they all have about them this quality of mathematical beauty. The use of the principle as a heuristic device for research in fundamental physics has impressive inductive support from the history of science. Nevertheless its invocation and use faces certain problems.

The first question is: How stable a guide is it? The investigation of the physical world with the longest history is humanity's attempt to understand the motions of the solar system. Throughout its two and a half thousand year history, the mathematical heuristic has been applied, but in rather different forms.[4] From Aristotle to Copernicus, it was the circle which seemed the acme of mathematical perfection. Accordingly, theories were formulated in those terms, though the increasing elaboration required considerably modified the elegance which might have seemed to have been achieved. That this programme was feasible at all is now seen as having depended upon an accident of the nature of the solar system, particularly that the eccentricities of planetary orbits are small, so that they are approximately circular. Kepler's introduction of ellipses must have seemed a blow to mathematical beauty (though he strove to reintroduce it in his own idiosyncratic way by his ingenious but mistaken use of the platonic solids). Mathematical beauty was truly restored when the insight of Newton established a universal law of gravitation based upon a particularly elegant law of force, that of the inverse square.[5] Geometrical elegance had given way to analytic elegance. Yet geometry returned to favor with Einstein's general relativity, based on the curvature of space and employing a specially simple formulation in terms of Riemannian geometry. If superstring theory succeeds, this picture will in turn be replaced by a much more abstract account in terms of the vibrations of one-dimensional entities.

All through this history, mathematical beauty has been there but its form has changed in a way that might make it look more like the fashionable than the fundamental. I think that judgement would be mistaken. From Newton onward, at least, when the picture changes the new paradigm is linked to the old by well-understood correspondence principles, explaining how the new theory and the old theory are related to each other in the domain in which they are compatible. These correspondences serve also to link the forms of mathematical simplicity, so that we can understand how the inverse square law and the use of the curvature tensor relate to each other. The arbitrariness which the epithet "fashionable" would connote is wholly missing.

This feeling is reinforced by considering another branch of fundamental physics, that concerned with the elementary constituents of matter. Although its modern history is only about two centuries old, it is intellectually a well-winnowed field, for it has gone through four stages of development in this time (from atoms to nuclei to nucleons to quarks), three of them in this century. In each case the pattern of change was the same: first an increasing complexity, followed by the recognition of patterns within that complexity, whose explanation was found to lie in the uncovering of a deeper level of structure, whose properties then restored an elegant parsimony of basic entities in a new form.[6] The new structures thus revealed have always found natural expression in compact mathematical terms.

The second question concerns the nature of mathematical beauty. What is it and can it be defined in a way that makes it more than the expression of an idiosyncratic preference? In the foregoing, I have repeatedly used words such as "economy", "elegance" and so on. It is clear that these point in the direction where the quality of beauty is to be found. However, no one has succeeded in encapsulating mathematical beauty in a definition, any more than its cousins in human aesthetic experience have proved to be codifiable in that way. Does this not, then, make it altogether too elusive a concept to bear the heuristic weight I am trying to put upon it? I do not think so. However hard, or impossible, it may be to define mathematical beauty, there is very considerable agreement among the mathematically aware about when it is present. We are in that area of tacit skill, so well described by Michael Polanyi,[7] in which we "know more than we can tell". I am content to accept the validity of this intuitive knowledge since I also accept Polanyi's analysis that such acts of personal judgement, taken within a competent community and pursued with universal intent, are fundamental to all scientific activity (not least in the exercise of the skill of induction).

We are left, then, with the fundamental question: How does it come about that mathematics is the key which unlocks the secrets of the structure of the physical universe? It is surely a non-trivial (a mathematician's phrase meaning "highly significant") fact about the world that this is so. After all, mathematics is conducted in the human mind. It is the exploration of a rational world within. Yet I am claiming that some of the most elegant patterns encountered in that interior voyage of discovery are found to be realised in the exterior structure of the physical world around us. The reason that we experience within and the reason that we observe without fit together in perfect consonance. This is the property of the "unreasonable effectiveness" of mathematics to which Wigner drew our attention in his essay. In his explicit discussion he lays less emphasis than I have done upon the *beauty* of the mathematics involved. I think that this is because the concept is implicit in Wigner's account of what mathematics is. He quotes with approval Polanyi's dictum that the most obvious feature of mathematics is "that it is interesting" and he goes on to say of the choice of axioms for a mathematical system that they "are defined with a view to permitting ingenious logical operations which appeal to our aesthetic sense both as operations and also in their results of great generality and simplicity".[8] Wigner calls the power of mathematics to give an account of nature "the empirical law of epistemology"[9] and he accords it the status of an article of faith for the theoretical physicist. The question is: Faith in what? Wigner prefixed his essay with an epigraph from C. S. Peirce: "and it is probable that there is some secret here that remains to be discovered".

One answer might be thought to lie in the effectiveness of the process of evolutionary biology. After all, our minds must display in their workings some considerable degree of conformity to the way things are if we are not to die of injury or starvation. Yet this can only explain the congruence of everyday thought (the mathematics of 1,2,3, ... and a little elementary geometry) and everyday experience (the macroscopic world of tables and chairs). What we are concerned with is something altogether different from that: very rarefied and abstract mathematics (Lie groups, Riemannian geometry) and physical regimes remote from the everyday – indeed often counterintuitive to its expectations (the quantum world, the vast universe discerned by cosmology). It seems incredible that Einstein's ability to conceive of general relativity, and so to succeed in formulating a theory of gravity by means of the simplest conceivable expression in terms of the curvature tensor, is just a spin-off from the struggle for survival. Wigner drily observes, concerning our impressive powers of abstract thought, that "certainly it is hard to believe that our reasoning power was brought, by Darwin's process of natural selection, to the perfection it seems to possess".[10]

Rather it would seem that there is evidence in nature of a deep rational structure to which our minds are attuned. That is scarely surprising, say those who follow Immanuel Kant, for the order you "discover" is in fact an order which you have imposed. The assertion is that our reason-seeking will moulds experience into shapes which are consequences of the way in which we are forced, by an *a priori* exigence, to cope with the flux of what is happening. The patterns perceived are necessary consequences of our epistemological procedures. Sir Arthur Eddington told the parable of the fishermen whose net has a mesh of 4″. They conclude that there are no fish in the sea shorter than four inches. Wigner appears to flirt with Kantian notions without finally commiting himself. He says, "We are in a position similar to that of a man who was provided with a bunch of keys and who, having to open several doors in succession, always hit on the right key at the first or second trial. He became skeptical concerning the uniqueness of coordination between keys and doors."[11]

There are two possible explanations of the man's success. It could be that the locks are so manipulatible that several keys would in fact work and he simply succeeds with the one he happened to like for some extrinsic reason. That would be the Kantian proposal. Alternatively, he might have been furnished with a powerful passkey which is capable of opening all the locks he encounters. It is the latter possibility which seems to me to be much the more likely interpretation of our successful exploration of the physical world.

If the reason we find were a reason we impose, this would imply a considerable degree of plasticity in our experience, permitting us to shape it to our fancy. The feel of physics research is entirely different. One of its most striking (and exciting) features is the way in which investigation of a new regime so often brings with it surprising discoveries. We do not find what, *a priori*, we might have expected. On the contrary, the physical world proves itself to be recalcitrant to our expectation. Rather than imposing our will upon nature, she appears resistant to our wishes (and, one might remark, to the forecasts of Kant and Eddington!). The difficulty is often for a while to find *any* theory which is both coherent in its structure and empirically adequate for a range of phenomena. Given the concentration of talent in the pursuit of fundamental physics, this can scarcely be attributed to the lack of active effort. The prolonged struggle which evolved modern quantum theory over the period 1900–1926 testifies to this fact of how difficult the physicist's task actually is. Yet the eventual theory, when found, proved compact and astonishingly powerful. Wigner rightly comments "Surely in this case we "got something out" of the equations that we did not put in".[12]

If the order that physics describes in nature is actually an order inherent in the phenomena, then how does it come about that our minds are so apt to discern it? It is a question which anyone imbued with the scientist's instinct to seek as full an understanding as possible, is bound to pursue. The most satisfying answer would surely be if the congruence of the reason within and the reason without could be ascribed to an underlying rationality which was the ground of both. Such an explanation would be provided by the theistic doctrine of creation. Of course, such an idea will seem most natural and attractive to those who, like myself, believe that there are also other grounds for a belief in God. Yet by itself, the role of mathematical beauty as the guide to understanding the physical world, is an encouragement to seek a Mind behind the order of that world.

The deeper rationality which links the reason within and the reason without is divine Reason; the signs of Mind with which the structure of the universe is shot through are signals of transcendence. We enjoy the intellectual pleasure of admiring the order of the cosmos because we are creatures living as part of God's creation, our rationality part of the divine image within us. The unreasonable effectiveness of mathematics receives its just interpretation through the insight of natural theology. I have written elsewhere

> it is to the appeal to intelligibility of the world that we must turn if we want the argument of natural theology with the greatest degree of fundamentality and the highest prospect of endurance. It is a fact about the world of manifest significance and proven staying power. Torrance says about the universe that science investigates that it "does have something to "say" to us, simply by being what it is, contingent *and* intelligible in its contingency, for that makes its lack of self-explanation inescapably problematic". He goes on to speak of the world's "mute cry for sufficient reason". St Augustine speaks of our hearts being restless till they find their rest in God. He had in mind principally the longing of love in the depths of our being, but it is also true that our intellectual restlessness will only find its final quiet in the vision of God.[13]

I believe that the role of theology is to provide the most profound and comprehensive setting within which to comprehend the autonomous conclusions of all other forms of human inquiry into the way things are. Theology is the ultimate expression of our deeply held conviction of the unity of knowledge. In the end, the search for an understanding through and through proves to be the search for God since, in Bernard Lonergan's phrase, he is "the unrestricted act of understanding, the eternal rapture glimpsed in every Archimedean cry of Eureka".[14] The congruence of the reason within and the reason without, experienced through the unreasonable effectiveness of mathematics, was described by Wigner as "a wonderful gift which we neither understand nor deserve".[15] I think we can understand it – as the gracious gift of our Creator.

References

1. See, for example, P. Davies, *Superforce* (Heinemann, 1984); H. Pagels, *Perfect Symmetry* (Michael Joseph, 1985).
2. E. P. Wigner, *The Unreasonable Effectiveness of Mathematics in the Natural Sciences, Commun. Pure Appl. Math.* **13** (1960) 1–14.
3. Quoted in M. Longair, *Theoretical Concepts in Physics* (Cambridge University Press, 1984), p. 7.
4. For a survey see, for example, D. Park, *The How and the Why – An Essay on the Origins and Development of Physical Theory* (Princeton University Press, 1988).
5. A particular beauty of an inverse square law is that it gives a constant flux through all surfaces.
6. See, for example, J. C. Polkinghorne, *The Particle Play* (W. H. Freeman, 1979).
7. M. Polanyi, *Personal Knowledge* (Routledge and Kegan Paul, 1958).
8. *Op. Cit.*, p.3.
9. *Ibid.*, p. 10.
10. *Ibid.*, p. 3.
11. *Ibid.*, p. 2.
12. *Ibid.*, p. 9.
13. J. Polkinghorne, *Science and Creation* (New Science Library, 1989), pp. 30–1. The quotation from Torrance is from: T. F. Torrance, *Reality and Scientific Theology* (Scottish Academic Press, 1985), p. 52. See also: J. Polkinghorne, *One World* (Princeton University Press, 1987), Chap. 5.
14. B. Lonergan, *Insight* (Longman, 1958), p. 684.
15. *Op. Cit.*, p. 14.

The Modelling Relation and Natural Law

Robert Rosen
Department of Physiology and Biophysics
Dalhousie University
Halifax, Nova Scotia
B3H 4H7, Canada

1. Introduction

When I was a graduate student in the Department of Mathematics at the University of Chicago, there was a well-known story told about Professor Andre Weil, and which I have since heard ascribed to a number of other mathematicians. According to the story, Weil was lecturing on an abstruse aspect of algebraic number theory. During the lecture, he wrote a relation on the blackboard, which he described as "self-evident". A moment later, he began to stare at this relation, lapsed into silence, and after a few moments left the classroom without a word. He repaired to his office, where he consulted a number of arcane texts. Some fifteen minutes later, he returned to the classroom with the remark, "Yes, I was right; it *is* self-evident."

I thought of this story, for the first time in years, while re-reading Wigner's (1967) lecture on the unreasonable effectiveness of mathematics in the natural sciences, in preparation for writing the essay which follows. Wigner would not have written his article if he felt that mathematics was only "reasonably effective", or even "reasonably ineffective" in science; the reasonable is not the stuff of miracles, and gives us no *reason* to reason about it further. Wigner obviously felt, however, that the role played by mathematics in the sciences was in some sense excessive, and it is this excess which he regarded as counter-intuitive. His essay provides evidence to bolster his impression, but he does not attempt to account for the excess which he perceives.

In what follows, I will attempt what Wigner did not; that is, I will try to make it reasonable that mathematics should be "unreasonably effective" in Wigner's sense. The reader can now, I hope, perceive why I was reminded of the little story about Andre Weil with which I opened.

There are two prongs to Wigner's disquiet about the role of mathematics in science. The first is that mathematics, taken in itself as an abstract entity, should have such success in dealing with extra-mathematical referents. The second, which builds upon the first, notes that the criteria which have guided the historical development of mathematics have apparently been unrelated to any such extra-mathematical referents; they have been, rather, criteria of (subjective) mathematical interest, internal beauty, and most of all, as an arena of the exercise of cleverness. Why, Wigner asks, should formalisms developed according to such criteria allow extra-mathematical referents at all, let alone with such fidelity?

Wigner's remarks thus bear strongly on the familiar distinction between "pure" and "applied" mathematics. This distinction rests entirely on the fact that the former claims to exclude all extra-mathematical referents, while the latter uses such referents in a rather decisive way. In other words, "applied mathematics" is about something, while "pure mathematics" claims not. Thus, Wigner's essay concerns the unreasonable ease and effectiveness with which pure mathematics can be transmuted into applied mathematics (but, the pure mathematician will say, not necessarily conversely).

2. Natural Systems and Formal Systems

Any observer, perceiver or cognizer automatically creates a dualism between himself and what we shall call, for want of a better word, his *ambience*. The ambience comprises all that inhabits the external or outer world, the world of events or phenomena (including other observers). We draw the sharpest possible distinction between ourselves and our ambiences, even though some of us, on extended further reflection, come to argue that the dualism is specious, and that either the observer or the ambience does not in fact exist. We shall discount these aberrations, mainly generated by psychologists, and proceed on the basis that there is a real, fundamental distinction between the subjective internal world of the observer, and the external world of events or phenomena.

Natural science, particularly physics, is one attempt to come to grips with what goes on in the external world. Indeed, for three centuries, since the time of Newton, physics has been the exemplar of what natural science should be; an ideal to be emulated, or better, a universal cover in which everything in the ambience is ultimately to be wrapped. This last is in fact the goal of what is called *Reductionism.* Thus, for the past three centuries, when one talks of the "philosophy of science", one is talking mostly about the philosophy of physics.

As we shall see, the science of physics has taught us to isolate conspicuous parts of our ambience, which we call *systems* (or better, *natural systems*). The extraction of a system from our ambience then creates a new dualism, between the *system* and its *environment.* As science has developed since the time of Newton, we have learned to characterize systems in terms of *states*, and to cast the basic problems of natural science in terms of temporal sequences of state transitions in systems. These sequences are determined in turn by the character of the system itself, and by the way it interacts with the environment.

We learn about systems, their states, and their state-transition sequences through observation; through measurement. This, of course, is what observers are geared to do. But a *mere* observer is not a scientist. Indeed, basic features of the observer's internal world enter, in an essential way, in turning an observer into a scientist. Often, these features are *imputed* to the ambience of the observer, and treated as if they too were the results of observation. But in fact, they are not.

It is apparent that the internal world of an observer is not described in the same terms as is his ambience, or better, as are the systems and environments with which he learns to populate it. Above all, we do not measure the internal world; we experience it. What we experience is not states and state-transition sequences; it is ideas, thoughts, impressions. The internal world is the seat of these, and of volition, intention and imagination.

A central role in this internal world is played by a most peculiar entity which we call *language*. In general, language is itself a dualistic entity. On the one hand, language is supposed to be *about* something; its whole point is to capture and convey meanings about things outside of itself. These things, that language can be about, may pertain not only to the external world, but to the internal one as well; to both at once, or even to neither. This basic function of language, to express extra-linguistic meaning, we shall call its *semantic* aspect.

On the other hand, language can be perceived as a thing in itself, with its own organization and structure; its own rules; independent of any meaning or other semantic features. We may call this its *syntactic* aspect.

Clearly, the two aspects of language, its semantic and its syntactic characteristics, are inter-dependent. Without an appropriate syntax, to govern the generation and manipulation of propositions in the language, meanings could not percolate as they do through arbitrarily complicated proposition sequences. Conversely, the syntactical rules themselves are in themselves evolved from the need for language to function effectively as a semantic vehicle.

We are going to take the view that a *formalism* is a "sublanguage" specified entirely through its syntax. The importance of making it a sublanguage resides in (a) its capacity, through the larger language in which it sits, for semantic function, and (b) we may explore a formalism through purely syntactic means, independent of any such semantic function, if we wish to do so. Because there is, as we shall see, a close analogy between extracting a formalism from a language, and extracting a natural system from an observer's ambience, we shall call a formalism a *formal system.*

In the broadest sense, *mathematics is the study of such formal systems.* In what follows, we shall be concerned in almost equal parts, with (a) the internal syntactic structures of formal systems, and (b) the ways in which semantic content can be attached to formal systems, and especially, with the close relations which exist between natural systems in the external world, and formal systems in the internal one.

3. Entailment in Natural and Formal Systems

One of the basic linguistic forms is the *interrogative.* In fact, the posing and answering of questions is one of the most characteristic of human activities. Despite this, it appears also that interrogatives constitute the least studied of all linguistic forms.

Indeed, the common heuristic use of the term "information" identifies it with "that which can be an answer to a question". This is, of course, a far cry from the "Information Theory" of Shannon, which measures "information" in discrete "bits", and which has nothing much to do with language at all.

Of all the different kinds of interrogatives which can be identified, we shall be most concerned with a single class, characterized by the appearance of the word "why?". The answer to this kind of interrogative is characteristically of the form "because...". More specifically: If the question is "why A?" and the response is "because B", then a relation is thereby asserted between A and B. This relation will be called *entailment*; B entails A.

Now we have already noted that language is a dual structure; on the one hand, we have a syntactic aspect pertaining to the structure of language *per se*, independent of extra-linguistic referents, and on the other hand we have semantic aspects, in which extra-linguistic referents (meanings) play the central role. There is thus a corresponding dualistic aspect to entailment, which we can likewise denote by *syntactic entailment* and *semantic entailment*, respectively. Clearly, the former kind of entailment depends only on internal linguistic rules for generating new propositions from given ones, independent of meanings or interpretations; syntactic entailment is thus identical to formal *inference* or *implication.* Semantic entailment, on the other hand, is quite different; it does not refer to inherent properties of language, but rather pertains to the character of what the language is about. In what follows, we shall be concerned with a particular kind of semantic entailment, which will be called *causality*.

In more detail, causality is the name we give to relations between events in the external world. The concept was officially introduced into epistemology by Aristotle, who endowed it with the central role in all of science. Indeed, Aristotle defined science as being concerned precisely with "the why of things", and he thus identified scientific understanding with the ability to answer the question "why?". The Aristotelian doctrine of causality, which held sway until very recently (roughly, until about 1650, the time of Newton) held that there were distinct and independent ways in which one could correctly answer the question "why?" about an external event or thing; each way necessary, and all together sufficient, for proper scientific understanding of that event or thing. Aristotle argued that there were in fact four ways to say "because", which he identified with the categories of causation; formal cause, efficient cause, material cause, and above all (for him), final cause. These ideas have not disappeared from science, though their form has been radically changed (cf. Rosen, 1985a).

We have argued above, in effect, that the central concept in the actual deployment of language for human purposes is entailment. We have identified two distinct kinds of entailment; a syntactic one, governed by internal inferential rules independent of any meanings or referents, and a semantic one, governed by relations between what the language is used to describe. When we use language to talk about the external world, and about the events therein, these entailment relations between events comprise causality. Roughly, then, syntactic entailment is the province of mathematics and logic; semantic entailment or causality is science.

The trick now is to try to establish some kind of relation between the two kinds of entailment, and hence between the internal and the external worlds.

4. Natural Law

What follows is, in some ways, a short paraphrase of Wigner's (*loc. cit.*) discussion of these matters, cast into the framework of the preceding sections.

As Wigner points out, what he calls Natural Law consists of two independent parts. The first of these comprises a belief, or faith, that what goes on in the external world is not entirely arbitrary or whimsical. Stated in positive terms, this is a belief that successions of events in that world are governed by definite relations; such relations are precisely what were called *causal* in the preceding section. Without such a belief, there could be no such thing as science, and probably no sanity either.

The second constituent of Natural Law is a belief that the causal relations between events can be grasped by the mind, articulated and expressed in language. This aspect of Natural Law, which seems innocent enough, is decisive for our purposes, since it posits a relation between the syntactic structure of a language and the character of its external referents, of a type we have not seen before. So far, we have only talked about entailment within language or formalisms (i.e., about implication or inference, which relate purely linguistic entities), or we have talked about entailment between events (i.e., about causal relations between things in the external world). *Natural Law, however, posits the existence of entailments between events in the external world, and linguistic expressions (propositions) about those events.* It thus posits a kind of *congruence* between implication (a purely syntactic feature of languages or formalisms) and causality (a purely semantic, extra-linguistic constituent of Natural Law).

On the face of it, there appears no reason to expect that purely syntactic operations (i.e., inferences on propositions about events) should in fact correspond to causal entailments between events in the external world. Most of the time, in fact, they do not. Wigner's miracle is that sometimes they do; if we choose our language carefully, and express external events in it in just the right way, the requisite homology appears between implication *in* the language, and causality in the world described *by* the language.

A relation between a language or formalism, and an extra-linguistic referent, which manifests such a congruence between syntactic implication within language and causality in its external referent, will be called a *modelling relation.* We shall next describe such relations more precisely, before investigating what they themselves entail.

5. Modelling Relations

The diagram of Fig. 1 encapsulates what we mean by a modelling relation between a natural system N and a formal system or formalism F. The crux of the matter lies in the arrows of the diagram, which we have labelled (1), (2), (3), and (4). We shall explain what these arrows connote, and then write down the fundamental condition which embodies the modelling relation itself.

The arrow (1) represents causal entailment in the natural system N.

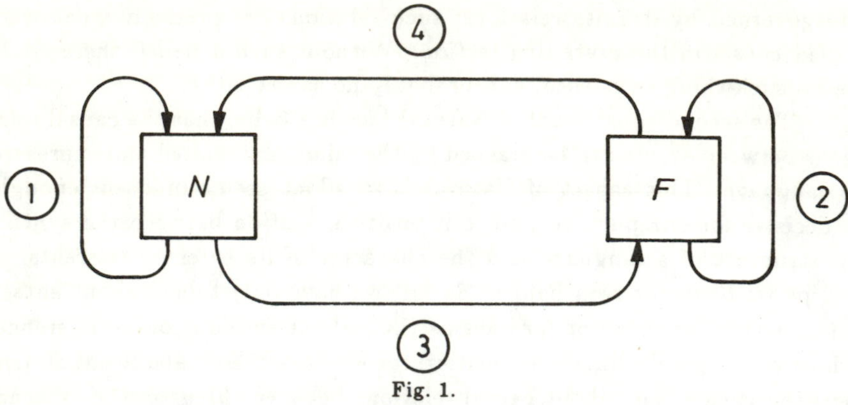

Fig. 1.

It may be thought of as the entailment of subsequent states by present or past states (although there is much more to it than this). It is what an observer seees diachronically when he looks at a system. The arrow (1) thus makes no reference to anything pertaining to language, or indeed to any internal activity of the observer, beyond the basic act of isolating the system in the first place.

The arrow (2) schematically represents the entailment apparatus of the formalism F; its inferential structure. This inferential structure is entirely linguistic; in fact, entirely syntactic. It makes no reference to semantics, meaning, or any external referents whatever.

The arrows (3) and (4) are the crucial elements. These arrows are what endow the various elements of the formalism F with specific external referents, and conversely, endow specific aspects of the natural system N with linguistic tokens in F.

The arrow (3) will be called the *encoding* arrow. It serves to associate features of N with linguistic counterparts in F. The simplest and perhaps the most familiar kind of encoding is the expression of results of measurement in numerical terms. Numbers, of course, are formal objects; the association of numbers with meter readings is the most elementary kind of encoding of a natural system in a formal one.

The arrow (4) denotes the complementary activity to encoding; namely, the *decoding* of elements of the formalism F into (observable) properties of the natural system N.

The arrows (3) and (4) taken together thus establish a kind of dictionary, which lets us pass effectively from the natural world to the formal one and back again. However, we may remark here on the peculiar status of the arrows (3) and (4). Namely, they are not a part of the formalism F, nor are they a part of the natural system N. These arrows are not entailed by anything, either in N, or in F. They cannot be meaningfully said to be *caused* by N or anything in it; nor can they be said to be *implied* by anything in F. And yet, as we shall now see, they are not entirely arbitrary either.

As we said earlier, a modelling relation between a natural system N and a formalism F involves a congruence between the causal structure of the former (the arrow (1) in Fig. 1) and the inferential structure of the latter (the arrow (2)). The vehicle for establishing a relation of any kind between N and F resides, of course, in the choice of encoding and decoding arrows; the arrows (3) and (4). The condition for congruence involves all these arrows, and may be set down schematically as follows:

$$(1) = (3) + (2) + (4) .$$

If this relation is satisfied, we shall then say that F is a model of N; or conversely, that N *is a realization of* F.

Let us explain more fully what the above congruence condition means. As we recall, the arrow (1) connotes the internal causal entailment structure in the natural system N. If we think of this entailment as embodied in state-transition sequences in N, it is what an observer sees when he simply watches events in N unfold.

The encoding arrow (3) pulls features on N into the formal system F. More precisely, it endows these features with formal images in F; images on which the inferential structure of F may operate.

We may think of these images as "hypotheses" or "premises" in F. The inferential machinery of F then specifies what these "hypotheses" entail within the formal system F; this process of entailment in F is precisely the arrow (2). The results of applying inferential rules to hypotheses generates "theorems" in F. The particular "theorems" in which we are now interested are those arising from hypotheses coming from N via the encoding (3).

It is evident that such "theorems" become assertions about N ("predictions"), when decoded from F to N via the arrow (4). The modelling condition then requires that we always get the same answer, whether we (a) simply watch the operation of causal entailment in the natural system N itself, or whether we (b) encode N into F via the arrow (3), apply the inferential machinery of F to what is encoded, and then decode the resulting theorems, via the decoding arrow (4), into predictions about N.

Stated yet another way, the modelling relation asserts the commutativity of the diagram in Fig. 1.

6. Some Ramifications of Modelling Relations

The deceptively simple diagram of Fig. 1 above allows us to reformulate the concept of Natural Law in a coherent way. Briefly, Natural Law asserts that any natural system N possesses a formal model F, or conversely, is a realization of a formalism F. Stated another way, Natural Law says that any process of causal entailment in the external world may be faithfully represented by homologous inferential structure in some formal system F, modulo the appropriate encodings and decodings. It must be stressed that Natural Law alone does not tell us how to accomplish any of this; it merely says that it can be done.

Thus, in a sense, the concept of Natural Law already entails the efficacy of mathematics, which Wigner appeared to find so unreasonable. Perhaps what is truly surprising is that we should have been so good at it. But that is another matter.

Modelling relations between natural and formal systems possess many other deep properties, which have not been paid sufficient attention as yet, and which have prospects for radically changing the manner in which the physicist (or more generally, the natural scientist) looks at the world. We will briefly mention a few of these.

6.1. Analogous Systems

Let us consider the commutative diagram shown in Fig. 2. In this diagram, we have two different natural systems N_1, N_2 which possess the same formal model F (or alternatively, which constitute distinct realizations of F). It is not hard to show that we can then "encode" the features of N_1 into corresponding features of N_2 and conversely, in such a way that the two causal structures, in the two *natural systems* N_1 and N_2, are brought into congruence. That is, we can construct from Fig. 2 a commutative diagram of the form of Fig. 3. This looks exactly like a modelling relation, except that *it now relates two natural systems*, instead of a natural system and a formal one.

Under these circumstances, we may say that the natural systems N_1, N_2 are *analogous*. Analogous systems clearly allow us to learn about

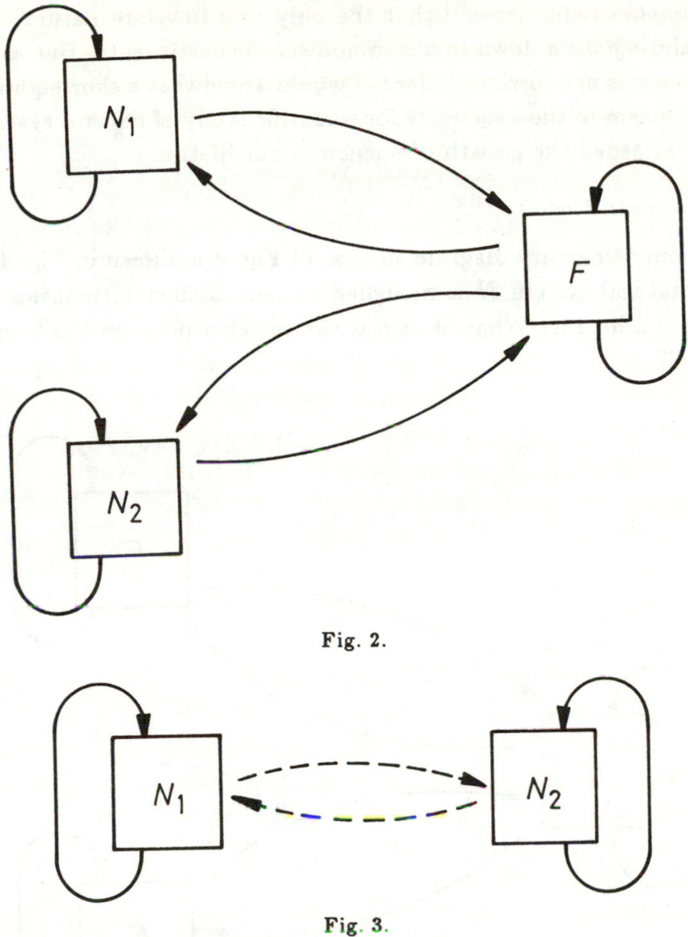

Fig. 2.

Fig. 3.

one of them by observing the other. Relations of analogy underlie the efficacy of "scale models" in engineering, as well as all of the various "principles of equivalence" in physics. But the relation of analogy cuts much deeper than this. Clearly, natural systems of the most diverse kinds (e.g. organisms and societies, economic systems and metabolisms) may be analogous; analogy is a relation between natural systems which arises through their models, and not directly from their material structures. As such, analogy and its cognates offer a most powerful and physically sound alternative to *reductionism*. It is often asserted, and still more widely believed, that physics *implies* reductionism; that the only way to relate natural systems is by analyzing them down to a common set of constituents. But as we can see, this view is not correct. Indeed, I would argue that a shortsighted faith in reductionism as the *only* mode for scientific study of natural systems has seriously retarded the growth of science in our lifetimes.

6.2. Alternate Models

A complementary diagram to that of Fig. 2 is shown in Fig. 4. Here, a single natural system N is modelled in two distinct formalisms F_1, F_2. The question here is: What, if any, is the relation between the formalisms F_1 and F_2?

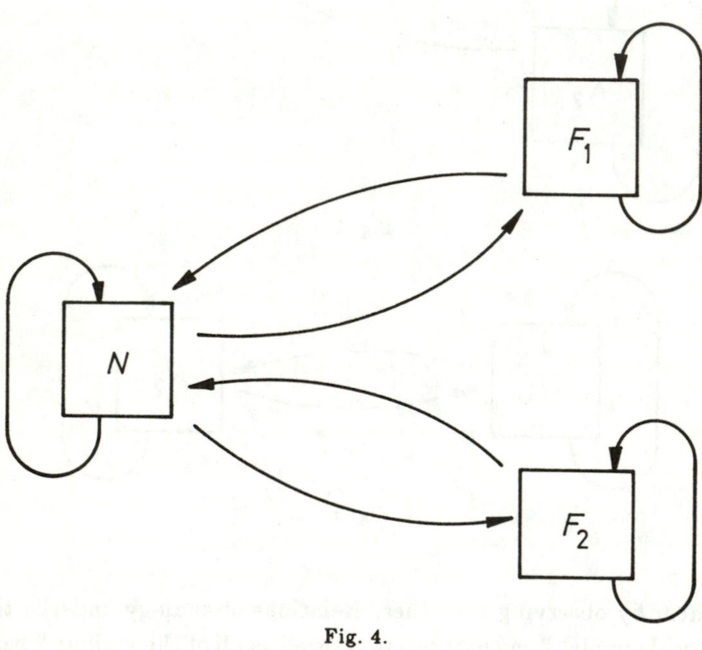

Fig. 4.

The answer here is not in general as straightforward as before; it depends entirely on the extent of the "overlap" between the two encodings of N in F_1 and F_2, respectively. In some cases, we can effectively build at least some encoding and decoding arrows between the two *formalisms*; a well-known instance in physics is the relation between the thermodynamic and statistical-mechanical models of fluids. In other cases, there exists no formal relation between F_1 and F_2; we then have the situation in which N *simultaneously* realizes two distinct and independent formalisms.

A host of practical and theoretical questions are raised in situations of this last type. Some of them bear crucially on the limits of reductionism itself. We have already noted that reductionism embodies the idea that the only way to relate natural systems is to analyze them down to a common set of constituents; molecules, or atoms, or elementary particles. The end result of such an analysis is an encoding of any natural system into a formalism which serves to express any system property in terms of these ultimate constituents. In some sense, this is the largest formalism, the largest model, which can exist; any other model is, in formal terms, some kind of quotient or submodel of this biggest one.

In this case, the independence of two formalisms F_1 and F_2, which we suppose that N simultaneously realizes, is only apparent; reductionism holds that we can always embed F_1 and F_2 in some larger formalism F, which is again a model, and from which F_1 and F_2 can be recaptured by purely formal means.

The existence of such a largest formalism, which itself models a given natural system N, and from which all others can be formally generated, would constitute a new postulate about the nature of the material world itself. I do not believe that such a postulate is valid. It seems that some kinds of natural systems admit such a largest model, but that others do not. Indeed, if we pursue this matter further, it appears that the distinction between those which do admit a largest model, and those which do not, has many of the properties of the distinction between inanimate and animate, or of simple and complex. This is not the place to pursue these matters; the interested reader may consult Rosen (1985b) for a fuller discussion.

It is at any rate apparent, I hope, that the concept of Natural Law, particularly as embodied in the Modelling Relation, has many deep ramifications; some of these bear directly on the innermost nature of the scientific enterprise itself.

7. A Postscript: The Effectiveness of Mathematics in Mathematics

Throughout the foregoing discussion, we have been concerned with making manifest some of the parallels which exist between the external world of events or phenomena, and the internal world of language and mathematics. Indeed, the modelling relation, and the concept of Natural

Law on which it rests, is merely a direct expression of these parallels. In this section, we will suggest some further extensions, in which the parallels themselves become objects of study.

It is perhaps not generally realized how much modelling goes on *within mathematics itself*. By "modelling within mathematics", I mean the establishment of homologies between different kinds of inferential structures, arising from different parts of mathematics. In effect, one part of mathematics is treated like the external world in science; its inferential properties treated like causal entailment. Such "internal modelling" within mathematics allows us to bring one part of mathematics to bear on another part, often to the most profound effect.

Examples of such "internal modelling" abound. Consider, for instance, Cartesian Analytic Geometry, which created an arithmetic model of Euclid's *Elements*, and thereby brought algebraic reasoning to bear on the corpus of geometry. Later, the consistency of "non-Euclidean" geometries was proved by establishing Euclidean models of non-Euclidean objects. Whole theories in mathematics, such as the theory of Group Representations, rest entirely on such notions.

Poincaré ushered a new era into mathematics by showing how to build other kinds of algebraic models of geometric objects. His idea of homotopy, and later, of homology, showed how to create group-theoretic images of topological spaces, and to deduce properties of the latter from those of the former.

In 1945, a whole new branch of mathematics was developed, initially to formalize these methodologies initiated by Poincaré. This came to be called the Theory of Categories (cf. Eilenberg and MacLane, 1945), and its subject matter was precisely the relations between different inferential structures within mathematics itself. In fact, it can be regarded as a *general theory of modelling relations* within mathematics.

Space precludes going into these matters in depth here; we must refer the reader to one or another of the excellent texts which now exist (e.g. Mitchell, 1965; Arbib and Manes, 1975). Suffice it to say that in this theory, the active agents of comparison between categories (i.e., between different kinds of formalisms or inferential structures) are called functors. The formal counterpart of Natural Law in this purely abstract setting is the existence of nontrivial functors between categories. Functors themselves can be compared, through the agency of *natural transformations*; these provide an agency of "comparing comparisons", of relating two different kinds of modelling procedures (cf. Fig. 4). This whole process can be iterated; natural transformations treated themselves as "hyper-functors" acting on a new category of functors, and so on.

The Theory of Categories is thus a formal image of the modelling process itself; not just of some specific way of making models of one kind of thing in another. It thus generates mathematical counterparts of epistemologies, but entirely within the formal realm, out in the open where they can be investigated in ways which are closed off from us in the external world.

Let us, for example, look briefly at the original motivation for developing Category Theory; Algebraic Topology. The basic problem of Algebraic Topology is a classification problem; to be able to tell whether two given topological spaces are identical (homeomorphic) or not. The group-theoretic models of Poincaré provide partial answers to this question. These models generate numbers (group-theoretic invariants), which are very much like observables of the associated topological space; if these numbers are different for two given spaces, then the spaces cannot be homeomorphic. The basic question in algebraic topology is whether there are "enough" models to discriminate any two topological spaces in some kind of effective way.

The question is still open. But it appears that sometimes there are "enough" such models, and sometimes there are not. When there are, we can build a "largest" algebraic model of each of the spaces involved, and settle the classification question by comparing these "largest" models. When not, then there is in effect no set of invariants, no set of "observables" whose values settle the question. It will be seen that this is a kind of abstract image or encoding of the questions we raised in the preceding section; and whose answers will have the most profound effect on physics itself.

Finally, we should point out that there is a whole "Theory of Models", which has to do with axiomatics and foundation studies. This has to do mostly with foundation studies; with axiom systems. It raises some new questions, because it takes a very particular view of language in general, and of formalisms in particular. We can only just touch on them here.

Basically, these questions arise from the extremely impoverished view of syntax which characterizes foundational studies in mathematics. The basic idea is that a formal system should consist of nothing more than a finite set of meaningless symbols (alphabet symbols) which can be combined according to a finite set of syntactical rules (production rules). Thus, formal language (including mathematics) becomes in effect a game of symbol processing, a game which can be played by a machine (e.g. a Turing machine). "Effective" processes are in fact identified with algorithms that can be executed by such machines. These ideas go back to David Hilbert, who clearly hoped to solve consistency problems by recasting all of mathematics in this way; he believed that all of mathematics could be *formalized* (i.e., expressed in this kind of impoverished syntactical form) *without any loss.*

We may note that there is an obvious parallel between Hilbertian formalization, in which meaningless symbols are shuffled around by specific syntactical rules, and atomism in physics. In the latter, "meaningless" (i.e., structureless) particles are shuffled by specific impressed forces. Just as Hilbert believed that every aspect of mathematics could be formalized, atomism (in effect, reductionism) holds that every facet of material reality can be, and should be, and must be, reduced to the appropriate atoms, and the forces which move them.

It is well-known that Hilbertian Formalism was killed by Gödel (1931), who showed that Number Theory was already unformalizable in the Hilbert sense. In effect, Gödel showed that either (a) the syntax of number theory, and hence of most of mathematics itself, is too rich to be expressed in terms of brute atomization, or (b) the *semantic* content of Number Theory (i.e., the fact that it is *about* something) resists atomization, or (c) both.

The importance of the above for physics, or for natural science, lies in the fact that the syntactical limitations of mathematical machines are often imputed to, or extrapolated to, the representation of events in the material world. These limitations are expressed in terms of *computability*, and the upshot is the requirement that material systems cannot do uncomputable things; cannot behave in ways which cannot be simulated. This imposes, in turn, profound limitations on what can sit in the right-hand box of Fig. 1, and hence, equally profound limitations on the left-hand side (which reflect, of course, the causal entailments which can occur in the material world).

At first sight, this restriction does not look too terrible to the modern physicist; after all, his formalisms generally take the form of differential equations, which are always simulable. But what if we are dealing with a natural system N which has no largest model? I would like to conclude by suggesting that biology is full of natural systems of this type, and that coping with them requires the deployment of models whose syntactic structure goes as far beyond the limitations of contemporary physics as the phenomena of atomic spectra and chemical bonding transcended the physics of the 19th century. In that sense, I would suggest that the true effectiveness of mathematics in dealing with material reality has hardly begun to be explored.

References

1. M. A. Arbib and E. Manes, *Arrows, Structures and Functors* (Academic Press, 1975).
2. S. Eilenberg and S. MacLane, *Trans. Am. Math. Soc.* **58** (1945) 231–294.
3. K. Gödel, *Monatshefte für Math. Phys.* **38** (1931) 173–198.
4. B. Mitchell, *Theory of Categories* (Academic Press, 1965).
5. In *Theoretical Biology and Complexity*, ed. R. Rosen (Academic Press, 1985a).
6. R. Rosen, *Anticipatory Systems* (Pergamon Press, 1985b).
7. E. P. Wigner, *Symmetries and Reflections* (University of Indiana Press, 1967), pp. 222–237.

Structure and Effectiveness

La Verne Shelton
Department of Philosophy
University of Wisconsin
Madison, WI 53706, USA

Mathematics, which has been called "unreasonably effective", sometimes develops on its own, with problems and goals stemming from its own subject matter, and sometimes develops in tandem with other disciplines. In the latter case, mathematical structures may be articulated in response to the needs of other theories in natural science or may actually reveal directions in which the development of physical theories should proceed. No one doubts that such ventures – at least since the "scientific revolution" that began at the end of the Renaissance – have been successful. In fact, the spurt in the development of mathematics around this time is sometimes said to have made this revolution possible.

Mathematics is a science about abstract structures. I shall argue that, partly because most mathematical structures are more abstract than the structures that are the subject matter of the physical, biological, and social sciences, a Realist about mathematical structures should *expect* mathematics to be remarkably useful in advancing the sciences.

1. It is a truism that scientific disciplines have often met with success: geneticists have been successful in articulating DNA structure and functioning; earth scientists have been successful in corroborating theories of continental drift; and surgeons have been, to some degree, successful with open heart surgery. But the success of mathematicians might seem to have surpassed, by far, that of other scientists.

Note first that although my initial observation concerned the effectiveness of one *discipline* in advancing other disciplines, the three examples given in the last paragraph are of *people* or groups of people experiencing success in accomplishing their goals. Effectiveness (except in certain technical senses) is basically a property of human activities and is, in part, a measure of the degree to which people have achieved their personal goals. Notions of the effectiveness or success of *things* are derivative.

To say that someone's success is "unreasonable" would be to say that the person's achievements are far beyond what would have been predicted upon a "reasonable" assessment of the situation. For example, my goal may be to win a certain race. But a reasonable assessment of the situation – I am a very slow runner and others participating in the race are faster – would lead a rational person to the prediction that I will not win, but will finish "with the pack." Among those participating, I will give a performance that is near the bottom of what is average. However, it so happens that all of the better runners travel together to the race, are delayed by the washing out of a bridge, and arrive too late to participate in the race. *As it happens* I am the best of those who actually participate and I win the race. I am not only successful but unreasonably so. My success is mostly a matter of circumstances that were (i) unforeseen and (ii) were not deliberately caused by anyone immediately connected with the event; and (iii) I would not have been successful had these unforeseen and unintentional circumstances not occurred.

This notion of the unreasonable success of persons in achieving their goals, that is, the success that occurs in circumstances with properties (i), (ii), and (iii) above, is rough, but sufficient for these purposes. The underlying and key property, I believe, is that the person was successful for the "wrong reasons."

Effectiveness is a more complex notion. It presupposes success (at least some degree of success), but presupposes, as well, that the success is achieved in a certain manner. An effective teacher is not merely successful in getting her students to learn. As well, she does not waste time, does not instill hatred or resentment among her students, and so forth. For a teacher to be *unreasonably* effective would be for him to be effective even though his efforts are usually not of the sort that result in the kind of success that is accompanied by qualities that are not necessary for achieving success, but enhance that success; and because this effectiveness is not caused by those immediately connected with the events in question, or is caused by them, but not in the "usual" way. Again, some unforeseen circumstances bring about his effectiveness and the teacher is an effective one for the "wrong reasons."[a]

Let us move on to a derivative, impersonal notion of effectiveness. One might claim, for example, that a certain lotion is effective in preventing pimples. The personal goal – that of preventing pimples – is the second term of the relation. The first term of the relation is the process, device, etc. the person uses in achieving the goal in a felicitous manner. If the lotion is unreasonably effective in preventing pimples, then, given the reasonable expectations of the persons involved, the lotion ought not to have been very good at preventing pimples. Perhaps the lotion was just a sunscreen lotion – then (one would expect) it should not have the right ingredients for pre-

[a]Note that "unreasonable effectiveness" is somewhat more oxymoronic than "unreasonable success." We much more readily believe that an instance of success is merely fortuitous than that an instance of effectiveness is. Being effective seems to involve taking care to be effective and succeeding in being effective because of this taking care.

venting pimples. We have roughly the following situation; P (a procedure or process) is unreasonably effective in doing or achieving (respectively, an action or a goal) G if P is effective in doing G and, given what one believes about the "nature" of P (what its typical function is, what it contains as parts, etc.) and the nature of G, one would reasonably expect P *not* to be effective or as effective in doing or achieving G.

An alternative, impersonal notion – that P is unreasonably effective in doing or achieving G if P is effective in doing or achieving G and, given the "nature" of P and G, P ought *not* to have been effective in doing or achieving G – has the advantage of not depending on the possibly idiosyncratic beliefs of the person or persons observing the situation. The first notion would be more accurately represented as a three-place relation: P is unreasonably effective in doing G with respect to the beliefs of A (A being a person or group of persons). Evidently mathematics is unreasonably effective in assisting in the advance of science *according to the beliefs of many people*; but I would assert that it is not unreasonably effective in the second, non-indexed sense. Its effectiveness is reasonable given the nature of mathematics, the nature of science, and the nature of the employment of the former for the sake of the latter.

Mathematics is effective in that it enables natural and social scientists to partition events, objects, forces, fields, etc. into "natural kinds" that are relata of lawlike generalizations; to formulate hypotheses and organize phenomena in a manner fruitful for research; to render (general) conjectures more precise and therefore more testable; to give explanations for observed regularities; and to find new analogies and generalizations. Moreover, mathematics is reliable, whatever one's opinion about the nature of the grounding of mathematical hypotheses. Whether it is "empirical," "*a priori*," or something else, it is incontrovertible that corroborated mathematical assertions are much less likely to be falsified than are corroborated hypotheses in other sciences.

To say that the effectiveness of mathematics is unreasonable according to the beliefs of a certain group of people is to say that those concerned, upon taking stock of the situation and the nature of mathematics, ought not to have expected its application to be so felicitous. Now, in a certain sense, we all do expect, and have good reason to expect, mathematics not only to be effective in these endeavors but to be *as* effective as it is. That is, scientists seek for "mathematical models" of the phenomena they are studying just because they know the power of mathematics. Their expectations render any particular instance of the effectiveness of mathematics quite reasonable because it would be in line with what has happened frequently in the past. But this is a kind of expectation that is differently based than the phenomenon described in the previous paragraph. At a deeper level, many people believe that there is little explanation as to *why* the "right" mathematical theory should help along the endeavors of economists, physicists, or geologists so radically.

The expectations that make the effectiveness of mathematics *un*reasonable are not based on experience of the uses of mathematics and the inductive generalizations made from that experience, but seem rather to stem from beliefs about the *nature* of mathematics: it should not be the *kind of thing* to help science along in the way that it does. A look at another example (rather farfetched), in addition to a look at ways that one science might be effective in helping the advance of another, will make this barrier to understanding the effectiveness of mathematics clearer.

X usually stands on his head immediately before giving a concert. If X does not stand on his head before a concert, he plays less well than usual. After a number of positive instances of this, we accept the correlation and come to expect that standing on his head will be effective in making X play better. Our *expectation* is, under the circumstance, reasonable, but the circumstance — the effectiveness of X's standing on his head – is unreasonable. Unlike, say, knowing his music well and being in good mental and physical condition, standing on his head immediately before a concert is not the *kind of thing* that ought to help X play well.

That we often think that the effectiveness of one science in stimulating the advance of another is reasonable has to do with our inherent belief in the unity of science. The logical positivists were among the first to strongly advocate such a picture of the Unity of Science, though we have now discarded much of their point of view. (For example, we no longer think it would be desirable if all scientific theories had the form of mature theories of physics.) But, some of the less propagandizing aspects of this picture seem accurate. Science is unified in the sense that different sciences often study the same thing, but look at different aspects of that thing. To take an obvious example – physiology, psychology, ecology, and physics all make a study of human beings, in an important sense. In another sense, each of these sciences is about different things: physiology about the structure and function of cells in aggregates, psychology about relations among beliefs, desires, experience, and behaviour, ecology about the relationships among organisms and their environment, and physics about relationships among more basic constituents of reality.

We do not consider it unreasonable, however, when one of these sciences is helped along by advances in another or when a scientist in one field looks for a structure in another that will help to frame an explanation for some phenomenon in her own field. For example, physiologists analyse and determine the physiological workings of neurotransmitters and psychologists are able to use these biochemical results to make predictions about the effects of an imbalance of these neurotransmitters on a person's moods. Conversely, a person's mood turns out to be a parameter in the determination of the conditions of the tissues of that person's heart. Or the earth's magnetic field turns out to be the means by which some species of birds manage to return to the same place to mate each year – an example of the effectiveness of geology in advancing zoology.

I would maintain that when one "natural" science, such as neurophysiology, is effective in helping the advance of another, such as psychology, we do not consider this effectiveness to be unreasonable precisely because we believe that the disciplines have a common subject *matter*, even though they structure the "matter" differently or look at different aspects of it from each other. We would not be surprised – might even expect – that when a neurophysiologist looks at the nervous system with a certain theoretical lens, that some of what she determines would be relevant to that nervous system as it is articulated by the field of psychology.

In general, the more closely related two fields are, the less surprising are such "interpenetrations." The effectiveness of chemistry in advancing biology is expected both at the level following from inductive generalization (it has helped in the past and should continue to do so) and at the level that perceives the "nature" of each field (some branches of chemistry have the same subject *matter* as some branches of biology). The limiting case of this, of course, is when an advancement in one part of a field gives assistance in other parts of the field. For example, the cracking of the genetic code has changed research patterns in many of the biological sciences.

Further, we think an interpenetration is less surprising if a more abstract discipline is effective in advancing the research program of a less abstract one. That theories in quantum physics should have ramifications in chemistry is less surprising than the converse.

Below I will attempt to make this concept of "more abstract than" more precise. But first I would like to discuss why mathematics might not seem (by nature) to fit into this pattern of interpenetration that we consider natural for the rest of science. In sum, the reasoning is the following. While other sciences are thought to be about the physical (in a broad sense that includes social, psychological, ecological, etc.) world, mathematics is either thought to be about nothing, or about abstract entities that are independent of the physical world. Let us look at these (what I will argue are) misconceptions more closely.

2. There are many views about the "nature of mathematical entities." I wish to consider three major families of views: Realist, Anti-realist, and Creationist. For the sake of this discussion I suppose a Realist view of the *physical* world and of most of the entities apparently referred to by those making systematic study of this physical world.[b] From that standpoint, the Anti-realist and Creationist as well as a species of Realist are in worse positions for understanding the effectiveness of mathematics than is a species of Realist called the Structural Realist.

The Inadequacy of Creationism: Many have suggested that mathematical structures are "created" by those who wish to discuss them.[c] In contrast, physical objects are allegedly independent of our notice. The process of "creation" is that of making a description that is consistent and that uniquely picks out, "up to isomorphism,"[d] the structure to be discussed. This description is both necessary and sufficient for the existence of a mathematical structure.

[b] In all that follows I will use the word "physical" in a most general sense, which includes many things usually called physical, but, as well, the biological, the psychological, the ecological, the economic, etc.

My criterion is that a physical entity (event) (field) be in (take place in) (be a part of) space-time. Typically, a physical theory describes such entities, events, or fields. But, it could turn out that some things described by physical theory and, for this reason, usually thought of as physical, are in fact not physical, and that some things normally thought of as mathematical are physical.

[c] I will assume "Mathematical structures exist" is true and that statements about mathematical properties of mathematical structures have a truth value. However, the sense of the words "exist" and "true" (the sense can be as in "Sherlock Holmes exists") will vary according to the idiom under discussion. In contrast, I reserve the word "real" for the full Realist sense of existence.

[d] Ways of individuating structure that make a coarser partition than equivalence classes under the "is isomorphic to" relation are discussed in M. Resnik, "Mathematics as a Science of Patterns: Ontology and Reference," *Noûs* 15 (1981), pp. 529–549. For my present purposes the finer individuation criterion is preferable; but for a more detailed understanding of structure-matching, the coarser criteria are likely to be needed.

Even if this doctrine has some truth in it, most mathematical structures clearly are not created ex nihilo, in full glory and precision. Initially the mathematician may have in mind a vague description of the sort of structure desired. For example, a structure was once desired that modeled the naive notion of symmetry in a mathematically fruitful way. The use of the concept of rigid motions to describe symmetry led to the concept of a group: Some conjectures were made, problems solved, applications within and outside of mathematics thought of, theorems corroborated, and, using the results of this largely empirical procedure, a set of necessary and sufficient conditions for being a group – i.e., the group theory axioms – were devised.

Or an econometrician may seek to model a social phenomenon in a mathematical theory and may find no theory that will do the job satisfactorily. It may be possible to generate from the various unsatisfactory mathematical theories one that is a good model for the purposes at hand.

Or certain lawlike generalizations may be empirically determined initially in simple quantitative terms. A familiar example began with Kepler's laws of planetary motion: (1) that the planets move in planar elliptical orbits with the sun at one of the foci, (2) that their orbits describe sectors (with radius vectors from the sun to the planet) of equal areas in equal times, and (3) that the square of the period of the orbit is proportional to the length of the major axis of the planet's orbit around the sun. These were corroborated by observation but had not been explained. One way to explain laws is to deduce them from more fundamental laws; Newton sought this sort of explanation for Kepler's laws. But a more powerful mathematical method than any so far available – one that allowed one to represent, relate, and manipulate rates of change of motion – was required for this. The calculus, shakily framed in terms of infinitesimals (called fluxions) was the result. Yet it was more than a century before the calculus was given a rigorous formulated and more than two centuries before it was rigorously formalized in terms of infinitesimals.

This last description of the creation of mathematics contains some shifts of reference. There are techniques of integration and differentiation used by Newton, Leibniz, and by those who came after them that are still in use today (though there have also been some changes in techniques). Most of these techniques can be thought of as following from a set of axioms of a theory that we have termed the calculus. But the theory of Leibniz was different from that of Cauchy and that of Cauchy from that of Weierstrass.

Both theories of Leibniz and (somewhat less clearly so, since he does use the concept of a limit) Cauchy quantify over infinitesimals. That of Weierstrass does not and uses the familiar ϵ/δ definition of a limit. Clearly then, the various *theories* are distinct. Are they "about" distinct entities? Since the theory of Leibniz is apparently inconsistent, is it about anything at all? Or is Leibniz's theory best thought of as a precursor of Abraham Robinson's, Leibniz's being a theory that would be about a similar relational structure to that of Robinson's if it were "cleaned up" and the apparatus of modern symbolic logic were used? Or, to put the question the other way around. If the relational structure that underlies the calculus was invented or created, when did it come to exist?

The Ontological Platonist, who, as a type of Realist, asserts that mathematical structures, which are neither in space-time nor mental, are real whether or not they are instantiated as physical structures,[e] has the easier task here and can characterize Newton, Leibniz, and their followers as groping for mathematically acceptable descriptions of real, independently-existing structures. The Creationist must, rather implausibly, assert that new analytic tools come to exist with each significant change in theory.

The Constructivist is a species of Creationist. Unlike other Creationists, the Constructivist would say that putative mathematical entities need not even belong in the field of the property of being real – that is, one can no more truely say that they are not real than that they are. A mathematical entity is real if it has been constructed in the course of a mathematical proof (the process) that has followed certain rules. If mathematical entities and/or the evolving structures that organize them have this nature, it is very difficult to see how the connection they have with physical reality makes their effectiveness in fruitfully organizing physical reality likely. The sole criterion of mathematical realness is the process of constructing an entity. Except insofar as the mathematician is a physical being, this process seems not to be constrained by physical reality. For example, the reality that Kepler described is in the causal ancestry of Weierstrass's calculus building – in some very loose and indirect sense the latter's calculus was developed partly to help put physics on a firmer foundation. But the nature of the interaction of the physical reality with the created mathemati-

[e] The terminology ("Ontological Platonist," as well as "Methodological Platonist" below) is due to Michael Resnik. See *Frege and the Philosophy of Mathematics* (Ithaca: Cornell University Press, 1980).

cal reality (mediated by descriptions in the minds of Weierstrass and many others), the nature of the bridge that renders the created mathematical structure useful, is entirely unindicated.[f]

Some sort of Creationist criterion for success in creating a mathematical structure that makes it more likely that "acceptable" structures would be effective in advancing other sciences might afford the opportunity of explaining this effectiveness. Let us look at some possible constraints on successful creation.

The Union axiom (rather than some contrary of it) is "accepted" as a part of our set theory partly because it is in accord with a preconception that any two sets can be combined to form another set. We would not accept an axiom that asserted that in some cases such a union of sets was not a set – there could, intuitively, be nothing but farfetched reasons for such exceptions.

Consistency is required for acceptableness. It is true that a useful, inconsistent theory might be endured – as in the case of Newtonian and Leibnizian calculi. But that will generally be for the reason that no one yet sees a way of developing a consistent alternative that would be equally useful. Also, to endure is to accept in a Pickwickian sense only. Once one has an alternative, the old tool is thrown out. For example, when Frege's set theory was found to be inconsistent, the source of that inconsistency was easy to isolate and it was soon possible to come up with consistent alternatives that were also useful.

Clearly, if the job of a mathematical theory is to model phenomena in the physical world, a mathematical theory is only acceptable if it does that job. Even if, as in group theory, the intent is just to model abstractions, such as types of mappings that preserve certain properties and operations on the mappings, the model has to be true to these abstractions. In other words, it may be that only certain structures are acceptable, even when we are within the realm of "pure" mathematics and not constrained by properties of the physical world. To take another example, in examining various axioms asserting the existence of large cardinals, the set theorist is moved to acceptance or rejection by consequences, for the more intuitive set theoretical notions, of adding such an axiom to, e.g., the basic ZF set

[f]Presumably the skin that is prone to be pimply *interacts* with the suncreen. But only certain kinds of interactions are likely to facilitate the lotion's being effective in preventing pimples.

theory.

Or perhaps a new theory is intended to generalize an old one, or to unify theories, or to idealize them. Or perhaps the new theory is meant to prove the same theorems (and perhaps more) as an old theory, but in a more elegant manner, or a more perspicacious manner, or ....

In sum, mathematical theories are rarely developed in isolation from other theories and in isolation from other practice of mathematics. Each new theory, in order to be acceptable, must be connected to older ideas and theories. Sometimes the new theory connects only to mathematical theories, but often, at least ancestrally, it also connects to physical theories as well as to the social context of research.[g]

These seem to me to be representative of criteria beyond "constructed in the process of a mathematical proof" that might constrain the acceptableness of mathematical structures. But a crucial point is that they could constrain the acceptableness of *discovered* mathematical structures. That is, it is no less pertinent that a Realist about mathematical structures limit her subject matter to those structures that connect with other theories and structures in ways that have, in the past, made it more likely that the structures would be helpful in advancing science or mathematics. The Realist could, and does, consider just such connections as those alluded to above when deciding whether a piece of mathematical insight is worth pursuing.

At first, the fact that the Creationist could use just such criteria of acceptableness as the Realist does would seem to undermine the view that Realism can give us a vehicle for understanding the effectiveness of mathematics but Creationism cannot. For, an opponent might argue: The Creationist can have similar criteria of acceptableness to those of the Realist. The Realist's criteria can be used to explain the effectiveness of mathematics. Therefore, the Creationist's criteria of acceptableness can help explain this effectiveness.

However, the second premiss is wrong, or at least incomplete. Acceptableness criteria help with an explanation only if one can see in what way the mathematical structures are "appropriately connected" with the subject matter of the disciplines they are effective in advancing. It does not suffice just to say "generalizing has been effective in the past." That

[g]As an example of the last, consider that the work being done by an assistant professor will not typically include the development of mathematical theories that those who are empowered to give her/him tenure would find very uninteresting.

tells the mathematician what to do, but not why (or how) doing it works. The Creationist can note that it is typically possible to weave two mathematical theories together by subsuming them under some more general mathematical theory. So, the Creationist might be able to use criteria of acceptableness to explain how, in this case, one part of mathematics is effective in advancing another: The more general theory is about a substructure of what each of the specialized theories is about; if we prove a result in one of these special theories, we can apply the same proof to the other special theory by first translating it into the more general terms. But, by hypothesis, mathematical structure, for the Creationist, cannot be appropriately connected with the structures of the *physical sciences* since the latter are discovered while mathematical structures are created.

Even in the case where the mathematical structure is created for the express purpose of better understanding certain physical phenomena, no "appropriate connection" and, hence, no explanation is indicated. If I say that I am going to create a five-foot long stick to measure the distance between the earth and the moon and it turns out that, somehow or another the stick does this, one is not enlightened by being told that it is designed for the purpose. What is missing is a story, which might be structural or functional, that tells how the construction of the stick is such that when related in certain ways to other structures, it interacts with them in such a way as to produce the desired result: A ruler works to measure a table because it is rigid and straight, is marked with units of measure, and can be laid along the rigid and straight edge of the table as many times as needed to encompass the table's length. It is difficult to see, however, how measuring sticks, which generally have such structure, but are only five feet long, reasonably might be used to measure the distance between one astronomical body and another. To say that it is designed to do so is akin to saying "I said a magic spell."

How can the Creationist weave a mathematical theory together with a physical theory? The unifying theory cannot be a physical theory because its subject matter would be real in the Realist sense. The unifying theory cannot be a mathematical theory because *its* subject matter would merely have been created.

This line of argument – which basically claims that the Creationist cannot relate the mathematical and the physical as they need to be related because he asserts that they have radically different ontological standings – might be disputed on grounds similar to the following. A painter creates a painting. It was not in existence until she created it. But, once created, it can interact with "natural" things, such as stones, breezes, and sulphuric acid, just as easily as with other artifacts. In the same way, created mathematical structures can interact with the objects of physical science.

But, the painting is an organization of "natural" things – pigments, chemical vehicles, wood pulp, etc. that were actual (instantiated) before the painting was created. The sense in which artifacts are *in* the natural world is that they *consist of* natural things in some arrangement. In general, mathematical structures do not consist of already actual physical events, objects, etc. If the natural things already have a certain structure, the mathematician has not created that structure by proving theorems under certain constraints. Even if the natural things do not, but *can*, have the structure, the mathematician has not created that structure by proving theorems. A structure that things can have is a structure that is real (though it may not be actual).

The painting is not *identical to* an organization, either. Even if it were, the objection does not hold: The painter is not making an organization real; she is making it actual.

(Analogously, a species defined by a cluster of genetic structures as they would manifest themselves in a certain, actual environment with a given history is real, but is not necessarily actual – it may never have been instantiated in the universe.)

Here is a different way of putting my observation. If the mathematician creates the mathematical structure, that structure cannot become real or manifest, even as a mere possibility, in the physical world – unless it was already manifest in that world and then discovered to be there.[h] If the structure is not manifest in the physical world, the mathematical theory cannot be effective in furthering our understanding of the physical world.

What this means is that the Creationist has given us no indication of an explanation as to why adhering to criteria of acceptableness should make it likely that acceptable theories (or the structures they are about) are effective.

The Creationist would seem not to be able to account for the effectiveness of mathematics in the easier direction – how we develop mathematics

[h] Of course, because we are assuming a Realist view of the physical world, the Creationist does not have the option of claiming that the mathematician structures the physical world (in the sense of moving the world about so that it has the structure) according to her theory.

so that it is effective. There are also cases in which pre-existing mathematical structures are incorporated into new theories in the physical sciences and, because of this incorporation, the new theories are able to develop in a rapid and fruitful manner. Examples include the incorporation of tensor analysis into relativity theory and the use of Hilbert Space in quantum mechanics. These mathematical structures were created for some other reason. Thus, it is even harder to account for their newly-determined effectiveness in newly-developing physical theories.

My conclusion here is that, given a Realist view of the physical world, the Creationist view of mathematics leaves the effectiveness of mathematics in advancing physical science a mystery.

Anti-realism: The arguments that Anti-realists leave the effectiveness of mathematics inexplicable proceed along similar lines, and I will do no more here than to indicate their general direction.[i]

I distinguish two types of Anti-realists, the Nominalist and the Instrumentalist. Both consider statements that contain apparent reference to mathematical objects to be in need of a reinterpretation that eliminates that apparent reference. The difference between the two views lies in their account of what this apparent reference accomplishes. For the Nominalist, it is not only an ontological excess, done perhaps for convenience or brevity. The apparent reference ought to be eliminated because in some sense or another, talk about such entities cannot be well-grounded. For example, it might be thought that we have no cognitive access to abstract entities (if there are such things) so can never know anything about them.

The Instrumentalist thinks that it is a correct description of the practice of science that mathematical theories are used as tools. Mathematicians, it is alleged, are, at best, Methodological Platonists (the term is due to Resnik) who act as though mathematical entities are real so that they can build structures that may have mathematical interest or may be applicable. Other natural scientists take up the fictionalist discourse so that their own research will go more efficiently and smoothly.

The views of the Instrumentalist and the Nominalist also make magic

---

[i]Stewart Shapiro, in "Mathematics and Reality" (*Philosophy of Science*, 50 (1983), pp. 523–548), makes similar points about the inferiority of instrumentalism in accounting for the effectiveness of mathematics. The type of argument he gives echoes, to some extent similar arguments against the instrumentalism of science in general: that the effectiveness of science can be best explained only if terms for "theoretical entities" really refer. Otherwise the explanation of this effectiveness also sounds like a magic spell.

of the effectiveness of mathematics. We are given no basis for understanding how the use of mathematical concepts and theories should assist the development of science.

3. The Adequacy of Structural Realism: According to the modern Platonist, abstractions are self-generating and are entirely unaffected by physical reality. Physical things, however, "participate" in the abstract entities that are their forms and, in some mysterious way, resemble them. But, we need to know how it is that mathematical entities can have this effect if they "exist" in a separate realm and what it is that enables them to illuminate properties of the physical realm.

An alternative Realism, Structural Realism, provides a basis for explaining the effectiveness of mathematics. Mathematical theories determine relational structures. Physical science, for the most part, is about physical entities and processes. Mathematics helps to structure these entities and processes (an example will be discussed at greater length below). These structures exist as actual or merely possible *real* entities and are, at least, possible structures of physical events, processes, and states.[j]

Mathematical and physical sciences are connected in the same way that physical sciences are connected with each other. A mathematical description of the three-dimensional enfolding of the hemoglobin molecule is just as much about blood as is the description of the chemical sequence of the hemoglobin molecule. The one science looks at determinants of geometrical structure while the other looks at determinants of chemical structure. The structure of hemoglobin is a mathematical structure and is a model of (or, at least, an idealization of it is a model of) an accepted mathematical theory. An even stronger point holds: The mathematical system that enables a scientist to predict and explain the three-dimensional enfolding of complex molecules was about the structures those molecules have (or approximately have) even prior to the actual application of the mathematical structure to the molecular system.

To say that mathematical structures are not Real (or even to say that they are "independent of the physical world"), that only the (physical) entities that "have this structure" are real, is rather like saying that there is no

[j] In many cases we do not know the "interpretation function." That is, we do not know what aspects of the mathematical structure are aspects of actual or possible events, processes, etc. The process of application is the process of discovering a useful interpretation function.

such thing as the color red; there are only red objects. Whatever is gained in the way of nominalistically-prone economy (or even Platonistic certainty) is lost in the inability to see how the electromagnetic spectrum can do any useful work in structuring perceptual reality. For, the Instrumentalist or Nominalist (about properties) cannot say "the spectrum matches the structure of physical reality." For them, the first term of the relation is a fiction. This is not to say that the Realist must say that the color red has some existence on its own, independent of the color blue or of a range of wavelengths among the radio waves. Rather, red is a (vaguely-bordered) region of the structure that is the electromagnetic spectrum, which latter structure is a region in other structures that, as a whole, model (or approximately model) theories of physics and that, as a whole, approximate to the structure of the physical world.

Mathematical theories may be categorical – have only models that are isomorphic copies of each other, C-categorical (C a cardinal) – all models with domains of cardinality C are isomorphic, or may have nonisomorphic models with domains of some particular cardinality. (First-order arithmetic fails to be aleph-null-categorical, for example.) Some authors have suggested that categoricity failures present problems for Structural Realists: For many central mathematical theories, there is no such thing as *the* structure that models a theory. The problem is troublesome if the Structural Realist is such a slave to axiomatics as to say that the formalized theory and nothing else determines the subject matter of mathematics. But, alternative, more correct versions of Structuralism do not depend on an identification of "the" subject matter of a mathematical theory with "the" (up to isomorphism) model of the theory.

As seen above, "the" calculus went through several distinguishable phases, some in which inconsistent theories were used some, in which formal theories that we have every reason to believe consistent were used. Yet the "applications" of the calculus, though they have diversified over these two hundred-odd years, have retained a common core in methods of differentiation and integration of common functions. I would submit that there is an entity that can be spoken of as "the" structure of the calculus. In my opinion, because it most closely approximates the intuitions that lie behind both the original development of the calculus and the means to understanding how analysis really works even in the face of ϵ/δ formalizations, the formal system that is closest to determining the structure of the calculus is Robinson nonstandard analysis. But a decision as to *which* axiomatic system "comes closest" is not important for the present purposes.

Axiomatic theories have the function of giving corroboration to mathematical (including "applied" mathematical) practice. They are "foundationalist" in the sense that a means is given for deriving mathematical assertions from a smaller (or at least more briefly specifiable) set of simpler or more intuitive or more "basic" assertions. A theorem derived in a formal system that has been proven consistent or, as is the more frequent case, in a system that has been corroborated through much successful use, is on much stronger epistemic ground than one which has been informally derived from a set of not-precisely-specified hypotheses. But mathematical knowledge is no different from the rest of our knowledge. A foundationalist picture of knowledge in general is inadequate: either the "axioms" or "basic assertions" themselves need support or they are inadequate to generate all our knowledge. Just as our general knowledge is unlikely to be encompassed by any foundationalist system, many interesting mathematical structures may never be so encompassed.

Evidence that the structures that form the subject matter of mathematics are not simply the models of its axiomatized theories can be found in mathematical practice. It is a commonplace that many "proofs" published in mathematical journals turn out to be faulty. The usual ploy – and it is usually successful – is to slightly change the conceptual assumptions behind the alleged theorem so that the theorem can be correctly proven. The result the original "proof" failed to establish is not simply declared false.

This constitutes a picture of groping after a means of formalizing intuitions, not a foundationalist picture of using a set of assumptions hewn in stone to derive results.[k] Euclid's theory of geometry, for centuries resting on a set of axioms insufficient to derive its theorems, is a notable case of structure outrunning theory. We did not wait for Hilbert to come along to consider the structure of geometry to be a mathematically real one.

To attempt to give an account of how we determine the mathematical

---

[k] Though I am somewhat sympathetic to the claim that we "intuit" mathematical structures, I do not think, nor does what I say here presuppose, that there is some separate faculty of mathematical intuition that can, for example, tell us whether the continuum hypothesis is false. Rather, when mathematics is seen as a growing body of knowledge (not necessarily monolithic) that has tight connections within itself as well as to theories outside of pure mathematics, intuitions can be seen as developing from familiarity with relevant parts of all of these theories and their manifestations.

structures that axiomatic theories only approximately describe would take us too far afield. But, with respect to specifying the referent of its theoretical terms through theoretical change and in spite of the insufficiency of axiom system to determine reference, mathematics is in exactly the same situation as other sciences.[1]

Mathematics is about structures then. Some of these, or parts of some of these, structures are structures in the physical world (for example, that of the real numbers). Some are idealizations of structures of the physical world (for example, some geometry or another). Some are merely possible structures. These structures are all real in the same sense in which physical structures like the spatio-temporal continuum, the electromagnetic spectrum, the set of elements, or the structure of RNA are real. Any of these structures may or may not be actual (or they may or may not be idealizations of structures that are actual).

To the extent that various structures "match," to that extent will one structure be useful in elucidating another. I would like to illustrate with an example that I have previously mentioned.

Kepler's laws are a summary, in geometrical terms, of carefully made measurements.

(1) The planets move in planar elliptical orbits with the sun at one of the foci;

(2) Their orbits describe sectors (with radius vector from the sun to the planet) of equal areas in equal times;

(3) The square of the period of the orbit is proportional to the length of the major axis of the planet's orbit around the sun.

The structures of reality that are asserted here to pertain include a certain type of conic section, which is the orbit of any one of the planets in physical space, some arithmetical relations, the structure of a solar system (sun plus planets in orbit around it), and time.

Newton's structure is partly specified by his second law.

(4) $\mathbf{F} = \mathrm{ma}$,

---

[1]Many readers will think that because scientists have "causal connections" with the objects of physical, biological etc. theories, the problem of establishing the reference of descriptions insufficient for picking out their referents is easier in other sciences than in mathematics. I can only issue a promissory note here, but would assert that since the relata of causal connections have a high degree of indeterminacy, the causal connections between scientists and their subject matter are of no help in establishing reference.

where both force and acceleration are vectors having three spatial components.

Further Newton *defines* acceleration to be the second time derivative of the position vector **s**. This adds the structure of the elementary calculus – whose domain contains, among other things, the real numbers, as well as infinitely small (less than any real) and infinitely large (larger than any real) numbers – to the structures already accumulated.

If the force acting on an object is zero, so is its acceleration. So, the velocity, which is the first time derivative of **s**, is a constant. This is Newton's first law.

(5) A body not acted upon by an external force will not change its state of motion.

The inverse square law further specifies the structure by giving the magnitude of the force between two bodies.

(6) $F = G\, m_1 m_2 / r^2$.

Here G is the gravitational constant in suitable units, m_1 is the mass of one body (say the sun) and m_2 is the mass of the other (a planet), and r is the distance between the bodies (here the length of the major axis of the ellipse). (This is an example of the alleged structure's being an idealization: the remainder of the bodies in the solar system are ignored as the force acting on each planet is considered; and the treatment is of a two-body problem.)

With this apparatus, Kepler's laws can be *derived*.[m] For example, to show that the orbit is planar, we assume some initial force vector pointing from the sun to the planet and some initial velocity vector. If the initial velocity is towards the sun, the planets will fall into the sun. Since this is not the observed situation, we assume that the velocity vector is not pointed towards the sun.

Consider the plane containing the velocity and acceleration vectors for the planet's motion. By the second law, all motion must be in that plane. (If it were not, another force would have to be acting to change it.)

Newton has matched the structure of his calculus to the physical-geometric and arithmetic structure presupposed by Kepler's observations. Theoretical identifications are made – in this case, the pretheoretical notion of acceleration as a change of direction or speed is identified with a second

---

[m] For the complete derivation from which this example is taken, see Saunders Mac Lane, *Mathematics: Form and Function* (New York: Springer-Verlag, 1986), pp. 259–264.

derivative. This enables the two structures (Newton's and the structure of the phenomena observed by Kepler) to interact or interpenetrate. The interpenetration proves fruitful in that what can be deduced is testable, agrees with observation, and, although my truncated presentation of the example does not show this well, provides new concepts and relations for further investigation.

The following is an approximate notion of one structure's matching another.

Structure S matches structure T if

(1) There is a common language, L, that expresses both a theory that (approximately) specifies S and a theory that (approximately) specifies T.

(2) The union of the theory of S and the theory of T in L is consistent.

(3) Theoretical identifications can be made relating S and T in such a way that

  (a) deductive consequences are obtainable from T and the identifications that were not obtainable before the identifications were made,

  (b) these deductive consequences agree with observation,

  (c) these deductive consequences are nontrivial in the sense that they provide new information about actuality and/or suggest new questions for exploration.

(4) The union of the theory of S with that of T, together with the theoretical identifications of (3), suggest a new composite structure approximated by a consistent theory that extends T but that has greater predictive and explanatory power than the theory of S and the theory of T together.

We can apply the notion to the above example. Here T is the structure of the phenomena Kepler observed – the sightings and measurements he took and the conclusions drawn from them. These conclusions can be described by a combination of arithmatic and physical geometry. This theory constitutes the theory of T. The theory of S is the fragment of the calculus consisting of assumptions Newton made about differentiation and integration; while the theoretical identifications are constituted by the assumptions he made when applying these techniques to physical parameters, such as acceleration, and the relations among these various parameters (the second law is one such relation). Kepler's laws are one of many deductive consequences of the matching and (although I do not take the example that far) the new composite theory has explanatory and predictive powers far in excess of the immediate purpose, deriving Kepler's laws, for which it was put together. One immediate result was an account of the perturbations in the orbits of the planets.

What has been attempted in this sketch is a framework for understanding what happens when a particular theory (S) is effective in changing another (T) into a new theory of greater power than either of them has alone.

4. Abstraction: It is the central assertion of this paper that mathematics is best viewed as another scientific theory that approximately delineates structures in the same way that physics or biology or other natural sciences do. All of these structures that the various sciences are about can be structures or idealizations of structures of physical states and processes. When one of these structures matches another, the theory of the first structure should be effective in advancing a theory of the second by changing the latter into a more powerful theory.

A remaining question is the following. Why is mathematics *particularly* effective? That is, genetics can sometimes be effective in advancing, say, zoology, but the effectiveness of mathematics is ubiquitous and rich. Why the difference? At least a partial answer lies in the degree of abstraction found in mathematics. I have argued elsewhere[n] that it is mistaken to consider the distinction between mathematical structures and physical structures to be a radical one: In particular, the abstract and the concrete are not fruitfully considered as *kinds* in a theory of structures or of theories.

It is helpful to single out a binary relation that applies to pairs of theories (taken as approximately, at least, determining relational structures). The relation of *one theory's being at least as abstract as another* creates a partial ordering among theories. Informally, given two theories S and T, S is at least as abstract as T if the class of maximal consistent states of affairs with which S is consistent contains (as a subclass) the class of maximal consistent states of affairs with which T is consistent. (Here a (logically) consistent state of affairs (or rather its specification) is maximal if, for everything there is, either the specification says it has a given property, fails to have it, or neither has it nor fails to have it.) If S is at least as

[n] "The Abstract and the Concrete: How Much Difference Does this Distinction Mark?" invited paper given for the American Philosophical Association Meetings in December, 1980.

abstract as T and T is at least as abstract as S, then S and T are equally abstract. If S is at least as abstract as T but T is not at least as abstract as S, then S is more abstract than T.

At least some of the order structure here corresponds to the hierarchy set up by the old Unity of Science hypothesis. By and large, mathematical theories will be more abstract than theories of physics and these more abstract than theories of chemistry or biology. But, it is not clear that the social sciences fit in here in a linear order – that is, it is probably the case, given that some physiological structures are not psychologically possible and, as well, that some psychological structures are not physiologically possible, that physiology and psychology are incomparable with respect to their abstractness. A more complex notion, possibly one that can assign a degree of abstraction to each theory would be needed to rectify the many gaps that would be found with the given notion of "more abstract than."

Even among mathematical theories, there are differences in abstraction. In general, extensions of theories, because they narrow down the relational structures that will count as models, will be less abstract than the original theories. Foundational sorts of theories, such as set theory and category theory, will usually be more abstract than theories that model specific phenomena (mathematical or otherwise).

The consistent theories at the most abstract end of the spectrum have more (in the sense of "superset of") ways of being true than those at the least abstract end of the spectrum. Ecological theories specify properties of many types of organisms, of non-organic substances, and of their interrelations. These theories would be among the least abstract.

"Matching", as I have defined it, is not a symmetric relation. One theory may help advance another (because the structure it is "about" matches the structure the second theory is about) without the second's advancing the first – except in the sense that any advance of knowledge is an advance of every theory involved. The theory that is given new direction and evolves into a more powerful one because of the matching process is more likely to be the less abstract one. Zoology is concerned with just living animals (past, present, and future), but genetics is concerned with a large range of possible organisms consistent with the workings of the genetic materials of actual organisms. Thus, structures of genetics are more likely to match those of zoology and result in zoology's enrichment and growth in explanatory and predictive power than the reverse.

The reason for this asymmetry is simple: changes in the less abstract theory will still be consistent with the more abstract theory. So the *opportunity* exists to make the less abstract theory more powerful. Further, the more abstract theory can be used as a sort of ordering principle for the material in the less abstract theory. Genetic structure is an ordering principle for the phenomenon of species in a manner similar to that in which differentiation and integration are ordering principles for physical phenomena. This follows from the nature of the process of abstraction. All sentences that are formed by a noun followed by a verb are, at this level of description, of the same form. A theory that describes language at the level of the parts of speech – noun, verb, adjective, etc., is more abstract than one that describes language at the level of particular words. The former serves as an ordering principle of the latter and can enrich it.

Consideration of such examples corroborates the claim that mathematics is as effective as it is because it is more abstract than most physical theories. Its structures are not only able to match those of many theories, but match them in such a way as to strengthen the theories that describe them.

In the converse case, when mathematical theories are extended or modified in virtue of their usefulness for a physical theory, the less abstract theory is actually pushing the more abstract theory into a position to better order the less abstract one. For example, philosophers have found first-order logic useful in formalizing natural language arguments for analysis. But finding that logical properties of certain arguments (for example: S knows that p. Therefore, p.) could not be elucidated by first-order logic, they have extended these mathematical theories by adding epistemic operators or alethic modal operators or whatever is needed for the necessary formalization. Philosophy drove the logic. The new mathematical theories were *for* philosophy.

This brings us round the circle. New mathematics developed for philosophy is also likely to find new uses – just as is new mathematics developed for the sake of mathematics.[o] And, again, this is because of the degree of abstraction of the new theories, whatever their original purpose.

[o]This brings up an interesting, and relatively rare, case in which another discipline, in this instance philosophy, rather *directly* illuminates pure mathematics. One system of alethic modal logic can be interpreted as a mathematical system that is an intuitionistic logic.

Psychology and Mathematics

James T. Townsend
Department of Psychology
Indiana University
Bloomington, IN 47405, USA

and

Helena Kadlec
Department of Psychological Sciences
Purdue University
West Lafayette, IN 47907, USA

A general theme of the present volume is to be "why and how mathematics works to describe reality". These questions are perplexing enough even in the physical sciences where mathematics has been used with the greatest success. The employment of mathematics in the biological sciences is perhaps somewhat less pristine. But the issue of mathematics vis-a-vis psychology is likely a provocative one in fields with longer scientific traditions. There are no doubt those who even question whether a science of psychology can ever be developed, much less whether one is presently extant. Since we are experimental psychologists, we obviously believe that such a science is possible, and only a little space will be devoted to this question. Rather our main concern here will be with whether and to what extent mathematics can be used productively in psychology. Needless to say, our search is not for some arcane analogue of astrology, but a rigorous apparatus that will serve psychology as fruitfully, if not perhaps in the same way, as it has so gloriously benefitted physics and other credible sciences somewhat further down in the pecking order of rigor.

But to return momentarily to the more general question posed for this project. Presumably we can all agree that there is order and structure in the universe, however complex that order may sometimes turn out to be. One is reminded of a statement attributed to Albert Einstein, paraphrased to wit: "When I was 16 years of age I set myself the task of putting down the laws of the universe in simple mathematical terms. I now consider that my self-imposed goal was doomed. I realize that the laws of the universe must be described in complex mathematical terms." Nevertheless, there are similarities of structure that are quite general in nature and that to some extent may attempt to mimic one another – sometimes in hierarchy and sometimes more horizontally. There may be allusions here to self-repairing and self-reproducing automata as foreseen by von Neumann (1958) and all the happy growths from that line of investigation. In any event, humankind's mimicking of universal structure seems to take the form of our attempts to describing the order that appears to us through our sensory organs (along with the instruments of modern science), in part by way of mathematics. Other paths are, of course, religion, philosophy and so on.

One suspects that other intelligent civilizations would (have?) put together stridently different linguistically (mathematically, etc.) coded accounts of nature, especially if their physical construction differs radically from ours. It seems reasonable to assume, however, perhaps as an article of faith, that there would exist metamathematical mappings that would carry their quantitative descriptions into ours and vice versa; at least for those classes of phenomena that intersect both "observable" universes.

With this general, if rather terse, philosophical background in place, we move to the more specific goals in ths essay. The three specific questions that will be considered are: 1. Can a science of psychology be developed? 2. If so, should mathematics be involved? 3. If so, how can mathematics best be employed?

The reader may not be surprised at our answer to the first question, as this is written by psychologists. Because the primary focus here is to deal with mathematics and psychology, not much time will be spent on this question. Suffice it to say that we take it for granted that a human is basically a system, albeit a complicated one, and that it therefore can be studied. Our personal predilection is to study the psychology of the individual (as opposed, say, to the study of social interactions) by varying the input, observing the output and drawing inferences about the internal systemic structure and its functioning from those observations. Such inductive inferences can be fed back into theoretical structures which can then be used to derive new predictions about input-output behavior, and these subjected to empirical test. This, in turn, leads back to revision of the theory, and thus repeating the inferential cycle. The clear emphasis is on behavioral research but physiological underpinnings should be taken advantage of where possible in order to provide direction and information about the "hard wiring" as well as its "programming".

We envision a sort of psychological systems theory which would perhaps be a special instance of general systems theory. This theory would possess its own peculiar characteristics that differentiate it from, say, that associated with the theory of automata, but might overlap and borrow from such disciplines, as in the theory of identification of canonical classes of automata. Although other scientific psychologists, as opposed to those engaged in clinical practice, might use a somewhat different language to express their scientific outlook for psychology, it seems safe to describe the modal viewpoint as close to the above.

Nevertheless, there is still massive disagreement as to the specifics of a science of psychology, even within the field itself. A prime example of this was the recent discourse on "methods and theories in the experimental analysis of behavior" by the venerable Burris F. Skinner, the legendary behaviorist (1984). This paper was published in *Behavioral and Brain Sciences*, an interdisciplinary journal, along with commentaries by a number of interested scientists including the first author. A formidable diversity of views is encountered there, to put it mildly. The reader is referred to that article for further discussion of these and similar issues and more references.

Question number two, whether mathematics should be involved in a science of psychology, might spark a bit more contention than the first one. First, it should be said that almost (but not quite!) no one in psychology would argue against the use of statistics in psychology; such use being ordinarily to describe data characteristics or test a hypothesis of some sort. In contrast, however, the present focus is on the use of mathematics as it is used in physics, as mathematical structures to serve as theories of psychological behavior, when the proper correlative definitions are drawn between the mathematical and empirical terms. To be sure, a brief scan of the major experimental and theoretical journals of psychology should convince one of the general acceptability of mathematical modelling and theorization. In psychology's most pristine subdisciplines, namely *psychobiology* and *psychophysics*, it would probably be taken for granted that a certain degree of mathematics is required for the advancement of the field.

Parenthetically, psychobiology (an older, and in some ways more expressive cognomen is physiological psychology) is that area of psychology that investigates psychological phenomena (especially overt behavior) through the study of germane biological aspects of the organism; for instance, physiology, anatomy, and biochemistry. Psychophysics is the study and description of sensory processes and perception particularly through the measurement of psychological reactions to different quantitative levels of stimulation on various dimensions (e.g., loudness as a function of acoustic energy; color perception as a function of wavelength of the stimulating light).

However, even in other areas of experimental psychology which pride themselves on their rigor, historically there have been two factions opposed to mathematics in psychology. The one faction would claim that psychology is simply not a "mathematizable" field. The second would maintain that psychology is still at an early inductive stage of its development, a stage where inductive empiricism and experimentation with regard to qualitatively expressed hypotheses is appropriate. Later, when a large body of orderly data is available, perhaps more mathematical theory may be adopted. The latter view appears close to that of Skinner, mentioned above (Skinner, 1984).

Interestingly, it is not rare to meet a devotee of a hard science (physics, chemistry, etc.) whose belief in the applicability of mathematics to nature stops short of psychology. Often, the said devotee can accept the general line of thought in Freudian or Jungian psychoanalysis – branches of psychiatry are not always in the best of odor even in that field, much less academic psychology – yet abhor the attempt to numerically measure psychological characteristics or more especially to render psychological theory in mathematical formulae. Presumably, "psychic (wo)man" should remain ineffable from the point of view of ordinary mundane scientific method.

It may surprise many from the traditional sciences to learn that experimental psychology was largely instituted in the nineteenth century by physicists and physiologists. To be sure, their avenues of investigation were largely guided by the philosophy of the times; particularly the British Empiricism of Locke, Hume, Berkeley, James and John Mill and others. Nevertheless, their methods came straight from the laboratory and the standard mathematics, especially the calculus of Newton and Leibniz, although the probability notions of Laplace, the Bernoullis, and Gauss were also present virtually from the start. Such well-known nineteenth century scientists as Helmholtz, Weber, Fechner, and Mach were heavily involved in the emerging science of experimental psychology (see, e.g., the classic history of experimental psychology by E. G. Boring, 1957). As psychology evolved to have a life of its own, its practitioners began to more formally identify themselves with the field, rather than with their original discipline. Here we see the first real psychological laboratory established in Leipzig in 1879 by Wilhelm Wundt (as every student in introductory psychology must memorize). Wundt's original training was in physiology and medicine.

In answer to the second question then, our opinion is that mathematics should indeed be involved in psychological investigation, at the level of theory construction and testing. In fact we would, as have certain others, turn the argument that psychological phenomena are too complex for immediate mathematical application around on the proponents of that view. It is precisely because the phenomena are so complex that we must have mathematics. Even in relatively simple (one might suppose) areas of psychology, a reader of the literature can easily be led down the primrose path through verbal argument. The logic seems impeccable. However, when the psychological principles on which the theory are based are put into mathematical form, the stated predictions may fail to follow at all. Moreover, just as in other sciences, the predictions are often rendered more testable by being derived as mathematical propositions or theorems.

If the notion that mathematics can be helpful in psychological theory is tentatively accepted, the third question posed is how should it be employed to best advantage? Although mathematics has been employed in psychology for over a hundred years, Clark Hull may perhaps be credited with the first serious program to formally model a substantial body of (mostly) animal behavioral data concerned with learning and motivation. He believed that much of the theory presumably confirmed with laboratory rats could be generalized to human behavior (e.g., Hull, 1943, 1952). Originally trained in electrical engineering, Hull self-consciously attempted to implement the so-called hypothetico-deductive method, an approach adopted from his comprehension of Russell and Whitehead's *Principia.* His approach also seems to have borrowed much from the Vienna Circle logical positivists and Bridgman's operationalism, at least in its general philosophy.

Thus, Hull's idea was to provide a system of primitive undefined terms, succeeding definitions and then axioms and theorems. Next, the theory would be related to reality by offering correlating definitions from the primitive terms, or more realistically and more often, from later axioms or theorems to empirical objects or measurements. Now the theory was in a position to be tested by interpreting the remaining theorems as predictions, given the necessary initial conditions.

This approach is pretty much standard today although usually in a less self-conscious fashion. Actually, many of Hull's predictions and experiments may now seem more like elaborated curve fittings than true attempts at theory testing, but his efforts set a new standard at the time. Following his death in 1952, much psychological/mathematical theorizing has been more modest, but often more rigorous. Certain developments have been relatively ambitious, for instance Estes' stimulus sampling theory (see, e.g., Estes, 1950; 1959; 1962; Atkinson and Estes, 1963) and Luce's individual choice behavior (Luce, 1959) attempted formulations to describe a wide variety of learning situation or decision situations, respectively. However, at that time, most mathematical theorizing had developed models for fairly limited experimental paradigms.

In approaching the third question, one might anticipate that once one accepts the notion of the usefulness of mathematics in psychology, a uniform mode of application would be forthcoming. This is far from the case. However, the diversity of mathematics presently employed may also be viewed as beneficial at this stage of development, as psychology searches for the appropriate mathematical descriptions within its various domains.

One way to view the spectrum of approaches is along the type of mathematics used; for instance, discrete versus continuous, deterministic versus stochastic, geometric versus algebraic, and the like. Although that has some advantages, it seems less useful than a depiction along the dimension of "process". That is, the types of quantitative theorizing in psychology can be placed somewhere on a continuum representing the extent to which the theory details the hypothetical inner workings of the psychological processes. If a theory or model is low on the process dimension, it would tend to be high on the descriptive dimension; thus, "process" and "descriptive" tend to be polar opposites. What follows is a brief overview of the various areas in psychology which have fruitfully employed mathematics in their theorizing. Hopefully this will convince the reader that mathematics has been useful in psychology, and will continue to be so even as we search for new mathematical models to guide our enterprise. We make no attempt to treat each area with equal space, but references are provided for those interested in more detail.

We begin with the area of *foundational measurement*, since one of the important questions one may ask is how can we measure psychological attributes. Foundational measurement takes as its goal the establishment of a rigorous theory of the assignment of number to real world qualities, especially psychological qualities. Because the traditional measurement of physics, which almost always presupposes the existence of extensive measurements with its associated ratio scale, is often too strong for the behavioral sciences, the theorists in this area have discovered other weaker types of measurement (e.g., Krantz, Luce, Suppes, and Tversky, 1971; Suppes and Zinnes, 1963; Narens, 1985; Pfanzagl, 1968; Roberts, 1979; Stevens, 1946, 1951). Oversimplifying for the purpose of exposition, it may be said that physicists can rely on the existence of a unique origin or zero point on their scales; something that is rare in the behavioral sciences. Only the unit can typically be altered. This cannot be taken for granted in the behavioral sciences where, more often, an interval scale (in which both the zero point and unit are non-unique) or perhaps an ordinal scale (in which only the order of objects on the scale is unique) would be appropriate for measurement. Moreover, it seems fair to say that quite basic questions and some resolution to them, have arisen through such enterprises, with regard to the fundamental nature even of the so-called extensive (ratio scale) measurement case associated with physics (e.g., Luce and Narens, 1987).

The area of foundational measurement has been closely tied to psychophysics because the latter tries to measure the sensory and perceptual reactions of human beings to physical stimulation as in loudness, pitch, and brightness as unidimensional examples. Color perception is a classical example of a multidimensional measurement, or scaling, problem (e.g., Burns and Shepp, 1988; Krantz, 1974), and although it has usually been rather descriptive in nature, in some cases it has been related to more detailed process models of a psychological or physiological nature (e.g., DeValois and DeValois, 1975; Hurvich and Jameson, 1957; Livingstone and Hubel, 1988). As in other applications of mathematics to psychology (psychometrics has been a partial exception), it is closely aligned with a class of experimental techniques and associated data.

Psychometrics, which develops tests of various psychological attributes such as intelligence, aptitudes, psychological pathologies, personality attributes and so on, is relatively far towards the descriptive end of the continuum. Historically, this line perhaps began formally with Spearman's two-factor theory of intelligence (Spearman, 1904; 1927), and the method continues to enjoy considerable popularity in the social sciences. This was later developed into the general methodology of factor analysis (Thurstone, 1947).

Psychometrics was originally also very closely connected with measurement in psychophysics as well as with assessment of individual differences (e.g., Galton, 1883), but it later went its own way. Today there is, as most academics, clinicians and industrial psychologists are aware, a large industry based on various psychological tests. The way in which psychological attributes that appear on the psychometric scales are defined (typically implicitly rather than explicitly) have usually been rather static. Some attempts to point out certain problematic mathematical or methological aspects of psychometrics seem to have gone largely unheeded (see e.g. Schönemann, 1981). Also, until recently there has been little attempt to develop an experimental and theoretical underpinning that might offer independent and explicit definitional status and confirmatory plexus to these attributes (as notable exceptions see, e.g., Embretson, 1985; R. Sternberg, 1984). Without such underpinning, the concepts tend to have a somewhat circular demeanor, in the view of the present writers.

An area of numerical research with wide applications that is historically related to the above approaches, is that of *multidimensional scaling* (e.g., Torgerson, 1958). It attempts to employ structure within a data set, often ordinal in nature, to retrieve a plausible geometric configuration of points in a multidimensional space, associated with some metric, usually a member of the Minkowski family (e.g., Attneave, 1950; Eisler and Roskam, 1977; Carrol and Chang, 1970; Kruskal, 1964a, 1964b; Shepard, 1964). There has been some interchange between traditional psychometrics and multidimensional scaling but the latter has also been employed in an extremely wide diversity of other basic and applied settings.

In another domain, in the 1940's and 1950's the developments of cybernetics, theory of automata, mathematical communications and detection theory, statistical decision theory, and information and coding theory terrifically influenced the behavioral sciences by providing new analogies for the study of behavior. In particular, they helped lead the field back from a rather stark behaviorism (even in the modified neobehaviorism of Hull, Tolman, Spence and others), to a more centralized, cognitive perpective of psychological processes. Once again experimental psychologists became interested in what might be going on inside the cognizer – we could say, from a systemic point of view, not necessarily explicitly physiological. The 1950's saw a heavy spate of studies employing information theory developed by Shannon (Shannon, 1948; Shannon and Weaver, 1949) to study the number of bits of information processed in various tasks and other related concepts (e.g., Attneave, 1959). It was said, not without foundation, that psychology tended to confuse information theory as a measuring tool with a substantive theory, which it was not at least for psychology (see e.g. Luce, 1959).

Another important development of the 1950's was *mathematical learning theory* (e.g., Bush and Mosteller, 1955; Estes, 1950, 1959; Atkinson and Estes, 1963; S. Sternberg, 1963). The Bush and Mosteller (1955) line tended to be somewhat less process oriented than Estes' stimulus sampling theory (e.g., Estes, 1959), but both were heavily dependent on relatively simple first order (and usually linear) differential equations to represent learning dynamics. Elementary combinatorics also played a role, especially in the stimulus sampling theory. Although out of vogue for a number of years, mathematical learning theory continues to make an important contribution to the understanding of human cognition (e.g., Raaijmakers and Shiffrin, 1981; Riefer and Batchelder, 1988).

Almost coincident with the appearance of mathematical learning theory was the invention of *signal detection theory*, an amalgamation of principles of statistical decision theory and notions of signals and noise from electrical engineering and applied physics (Peterson and Birdsall, 1953). It was almost immediately applied to human psychophysics and perception (Tanner and Swets, 1954; see also Green and Swets, 1966). Indeed, the prime originators were composed of electrical engineers (Birdsall and Peterson) and psychologists (Tanner and Swets) at the University of Michigan.

The early 1960's found a new paradigm emerging, the advent of the so-called *information processing approach.* Past issues of *Perception & Psychophysics* give a realistic picture of the evolution of this approach. This journal has closely followed the research in the area of information processing, especially those research problems that deal with perception. Charles Eriksen, the editor of *Perception & Psychophysics* over much of the period since its inception, has also influenced the information processing approach through his research contributions. It should be noted that this approach does not confine itself to information theory (see e.g., N. H. Anderson, 1981), and in fact rarely uses it. To the extent that it can be defined at all, it has tended to take as a metaphor a von Neumann type of digital computer (e.g., Arbib, 1964; Newell and Simon, 1963; von Neumann, 1958), if not something more simplified (e.g., S. Sternberg, 1966; see Roediger, 1980, for a discussion of the role of metaphors in the study of memory and Massaro, 1986 for a general treatment). Thus, the human information processor was often viewed as a set of processors arrayed in a series.

It was typically supposed that information entered by way of sensory apparatus and proceeded through the various processors (which varied with the particular cognitive or perceptual task under study) in a strictly serial fashion. Strictly serial here means that each subtask carried out by a process was completed before the information (used in a nontechnical sense) was sent on to the successive processor. Various psychological data were adduced to support seriality (e.g., S. Sternberg, 1966; Smith, 1968) and experimental and statistical methodologies were developed that depended on the truth of seriality of the processors for their success (e.g., S. Sternberg, 1969; cf. Townsend, 1971a, 1972, 1974). Other studies provided support for parallelity (e.g., Egeth, 1966; Egeth, Jonides, and Wall, 1972; Murdock, 1971). More recently, these methodologies have been developed for more general classes of processors including parallel and more complex networks (Fisher and Goldstein, 1983; Schweickert, 1978, 1983; Schweickert and Townsend, 1989; Townsend, 1984; Townsend and Ashby, 1983; Townsend and Sckweickert, 1989).

Most of the work in the general domain of information processing has not been heavily mathematical, but where it has been, the primary tool has been elementary probability theory and stochastic processes and the main observable variables have been reaction time (e.g., McGill, 1963; Pachella, 1974) and accuracy measures (e.g., Townsend, 1971b, 1981). In more complicated systems, computer simulation and computation that was sufficiently complex to require computer calculation began to see more and more use. A seminal paper in the latter vein was presented by Atkinson and Shiffrin (1968), work that proposed an information process type of human memory model. While most investigations over the years have employed the information processing framework, many have relied on more qualitative and intuitive approaches to testing hypotheses rather than a strictly quantitative strategy or theory (e.g., Craik and Lockhart, 1972; Massaro, 1972; Posner, 1973).

The 1970's witnessed a growing amount of more rigorous attempts at mathematical modeling (e.g., Krueger, 1978; Link and Heath, 1975; Luce and Green, 1972; Ratcliff, 1978; Snodgrass, 1972, Theios, 1973; Theios, Smith, Haviland, Traupman and Moy, 1973; Townsend, 1974, 1976a, 1976b; Townsend and Ashby, 1978; Wickelgren, 1970), and at separating classes of mathematical models based on opposed psychological principles, through the derivation of experimental tasks by mathematical means (e.g., Pachella, Smith and Stanovich, 1978; Snodgrass and Townsend, 1983; S. Sternberg and Knoll, 1973; Yellott, 1971).

One topic that received a good deal of attention, both theoretical and empirical, was parallel versus serial processing of multi-element visual displays (see Schneider and Shiffrin, 1977, for an excellent review of this literature). Shortly after some influential papers appeared that argued for seriality (e.g., S. Sternberg, 1966), the author (Townsend, 1969, 1971a) and Atkinson, Holmgren, and Juola (1969) pointed out the existence of stochastic parallel models which could mimic the standard predictions of serial models. Over the last twenty years, we and our colleagues have striven to map out the regions of the model spaces where experimental discriminability was impossible; that is, where the models make identical predictions in experimental paradigms. Townsend (1976b) and Vorberg (1977) generalized the equivalence theorems to large classes of distributions. Much of this work and further theoretical developments, including results showing how parallel and serial mechanisms *could* be experimentally discriminated, are reported in Townsend and Ashby (1983) and Townsend (1990). Luce

(1986) reviews a great deal of the quantitative work on response times over the past thirty years or so. A terse example representative of how mathematical theorizing can be used to develop tests of opposing psychological concepts will be given at the end of the paper.

In the meantime the dominant paradigm in *cognitive psychology*, which had by dint of fashion become an umbrella term (and still is) for virtually anything in psychology which was not strictly behavioristic, clinical, social or directly physiological (and sometimes, even those!), had evolved an even more fervid infatuation with computer simulation, and models based on digital computer ways of doing mental things. Thus, the birth of "cognitive science". Certain pioneers in artificial intelligence in general and computer science in particular, such as Simon, Newell, McCarthy, Minsky, Feigenbaum, Selfridge and many others, had always been interested in problems also considered as content matter for psychology, for instance problem solving, pattern recognition, speech and language production and processing, and so on. Many artificial intelligence theorists would be (and are) pleased when their models appear to be plausible candidates for human thought, while psychologists have also ventured into artificial intelligence (e.g., Hunt, 1975). However, when psychologists such as J. R. Anderson and Bower (J. R. Anderson, 1983; Anderson and Bower, 1973) began to get heavily involved in this type of work, experimental verification with human subjects became more of a daily integral part of such theorizing. Norman and Rumelhart and their colleagues (Norman and Rumelhart, 1975) were also prolific contributors to this literature with an emphasis on psychological realism and laboratory and natural data. To jump to the present, perhaps the most ambitious project now in the offing in this general sphere is SOAR developed by Allen Newell with the aid of many top rate experts in artificial intelligence and computer science (Newell, 1988).

Inevitably there have been many facets of cognitive science *a la* computer models that have been almost tropistic to philosophers and philosophers of science. This cross-disciplinary mixing of computer scientists, physiologists, psychologists and philosophers has served up a fascinating brew of theory and sometimes experiment, and in some cases, pure invective! An intriguing theory has been developed by Lefebvre (1982, 1987; see also Townsend, 1983) which is inherently psychological, yet quite unconventional with a strong dose of concepts not typically considered in good odor in psychology (i.e., sufficiently reducible to experiment, etc.). Nevertheless, he has made strides in mathematically defining such notions as conscience, ego, and good versus evil, and he is beginning to achieve a degree of experimental testing of his predictions.

One region of applied science with psychological implications is that of *decision making*. It has always (at least three hundred years!) had close ties with probability and statistics. Statistical inference is, of course, simply decision making under specified conditions. The Bernoullis, Laplace, Gauss, and many others provided the underpinnings of present day treatments, as later developed by Pearson, Fisher, Neyman, Savage and others. In the 1940's and 1950's, von Neumann and Morgenstern (1953) invented a brilliant axiomatic, and algebraic, machinery to deal with the concept of utility in uncertain environments. Their work has spawned an unbelievable number of papers and books over the past forty years or so. It seems to remain practically inviolate in the pertinent subfields of economics. Almost from the start, however, various of its axioms of predictions have evoked criticism from both economists (e.g., Allais, 1953; Ellsberg, 1961; Savage, 1954) and psychologists (e.g., Edwards, 1953, 1954, 1961, 1962) if interpreted as descriptors of human behavior.

Many modified theories, especially ones with less stringent conditions, have been proposed, especially by mathematical psychologists. A few of the latter are Luce (1959), Restle (1961), Fishburn (1976), Krantz (1967), Tversky (1967, 1969, 1972a, 1972b), Kahneman (Kahneman and Tversky, 1979; Tversky and Kahneman, 1981), and recently Goldstein (Goldstein and Einhorn, 1987), Miyamoto (Miyamoto, 1988; Miyamoto and Braker, 1988), and Busemeyer (1985), among others. Some, such as Coombs (1964) have developed alternative axiomatic approaches. All these latter approaches attempt to make the theories conform better to human propensities, including their decisional frailties!

Along the way, many nonquantitative experimental psychologists have been involved with proposing verbally based models and contributing data that have influenced the more mathematical endeavors (e.g., Payne, Bettman, and Johnson, 1988). Learning and motivational psychologists have also long been interested in decisions (e.g., Bower, 1959; Estes, 1976; Hull, 1952; Lewin (who was most of all a social psychologist), 1935, 1936). These investigators were inclined to take a more biological, or at least psychological, stance with regard to the background for decisions, with more attention to physiological substrates and environmental influences, especially those making themselves felt through learning and reinforcement (e.g., Hull, 1952; Tolman, 1959). Lewin (1935, 1936) observed that motivational variables often seem to act in a style suggestive of physical field theory, and this in turn was taken up by Hull (1943, 1952) and Miller (1959). However, the mathematical work on such a theory, which would be appropriate for psychology, was quite limited.

Recently, the first author and Jerome Busemeyer have begun work on a dynamic field theory for decision making, which mathematically attempts to describe real space-time behavior by people in decision making situations (Busemeyer, You, and Townsend, 1988; Townsend and Busemeyer, 1989; see also Nakagawa, 1987). One of the important aspects of decision making not captured by other theories is the "gravity-like" effect that positive attraction towards a choice object (e.g., candy) increases as a person (or animal for that matter) gets closer to it. Similarly, if the object also has unpleasant aspects associated with it (e.g., it makes one fat!), the negative attraction (i.e., repulsion) also becomes greater, the closer one is to it. It is fascinating, and well-studied with animals if less so with humans, that the slope of the negative attraction function, with the argument being distance from the object, tends to be steeper close to the object than is the slope of the repulsion function (e.g., Miller, 1959; but see also Hearst, 1969).

We would be remiss not to mention a major trend in experimental, particularly cognitive, psychology, namely *parallel distributed processing* (PDP; e.g., McClelland and Rumelhart, 1986; Rumelhart and McClelland, 1986). There are many features that help to define this area, some of them more critical and some of them quite fuzzy (as with any real-life definition). One of its most important aspects is that it helped psychology to self-consciously move away from the digital computer metaphor toward models that may work 'more like the brain', in some sense. The latter phrase is in single quotes because no one knows for sure how the brain works although much knowledge has been gained in the last 100 years. In particular, we still do not know for certain how memories are formed physiologically, though ideas abound; although the jury is still out, it seems that Hebb's (1949) idea that it is due to some kind of synaptic modification may not have been that far off the mark (e.g., Brown, Chapman, Kairiss, and Keenan, 1988).

Ironically, if naturally, memory is one of the favorite topics of the PDP approach. Within psychology, two major lines have evolved in modelling associative memory, both of which use vectors to represent various features of stimuli to be stored; one employs convolutions for storage and correlations for retrieval (e.g., Cavanagh, 1976; Eich, 1982; Murdock, 1982;

Pribram, Nuwer, and Baron, 1974; but see also Schönemann, 1987), while the other uses linear matrix operations (e.g., Anderson and Hinton, 1981; J. A. Anderson, 1973; Hintzman, 1986; Humphreys, Bain and Pike, 1989; Willshaw, 1971; Kohonen, 1977; see, e.g., also Pike, 1984; Willshaw, 1981). Nevertheless, we can say that PDP models seem to qualitatively, and perhaps globally, pick up some of the flavor of how the brain appears to operate more naturally than has been typical with models relying heavily on notions from digital computer land. For instance, they tend to be soft-fail, to go down as a continuous function of extent of damage (e.g., neural damage in real life), rather than in all-or-none fashion. Sometimes, they can act in a holographic manner, each part of the structure possessing almost all of the stored information; something that Lashley discovered to be approximately true for mammalian brain over fifty years ago (e.g., Lashley, 1929, 1933; see also Beach, Hebb, Morgan, and Nissen, 1960). As the name indicates, the operations are also often more parallel than serial, the latter being another holdover from early digital computer analogies.

Although the primary use of these "neural network" models in psychology has been in the modelling of human memory, they have also been extended to other and wide-ranging psychological and physiological phenomena. An extensive theoretical framework based on such neural nets has been developed by Grossberg and his co-workers, which, in addition to memory structures, attempts to account for such psychological properties as perception, motivation, and decision-making (e.g., Grossberg, 1980, 1988). Other extensions include the modelling of vision and pattern recognition (e.g., Fukushima, 1986; Koch, Poggio, and Torre, 1982; Marr, 1982; Poggio, Torre, and Koch, 1985) and various kinds of problem solving (e.g., Hopfield and Tank, 1986; see also Levine, 1983; Scott, 1977 for reviews). It could be said that this direction grew out of the "perceptron" line which has its roots in the early stages of artificial intelligence that was based on simplified physiological principles of the day (e.g., McCulloch and Pitts, 1943; Rosenblat, 1962), but was shown to be inadequate to compute relatively simple problems (Minsky and Pappert, 1969).

One of the problems that may plague this area, a problem that began to surface with the advent of computer simulation models, is that the models readily become so complicated that one could spend one's life striving to work out the laws of the functioning of the model. A number of investigators have voiced a concern about the testability of these neural net models (e.g., Ratcliff, 1981; Massaro, 1986; Schneider, 1987), and many consequences remain to be worked out. No doubt as with any new approach with promise in diverse areas, there will be much "noise" published. However, there is much that is intriguing with the work going on here and there is hope that rather a lot of "signal" will manifest itself as well.

Maybe it should be a capital offense to be writing about a partly mathematical subject without any observable mathematics. In order to avoid prosecution, we will limn in a little example of modelling in human information processing.

Consider the situation where a person is presented two objects to perceive, call them A and B, both shown at the same moment. If perception is relatively fast, it may not be easy to determine whether the person is capable of processing both objects simultaneously (in parallel) or must perceive them one at a time (i.e., serially). Suppose that if processing is parallel, it is statistically independent on the two objects. We do not assume that processing is equally fast on the two objects; that is, the distributions on processing time can differ. If processing is serial, we permit either object to be processed first, say A is first with probability p and B is first with probability $1 - p$. Now, it turns out that in order to have the serial and parallel models equivalent here, it is basically necessary and sufficient for them to be equivalent on the minimum processing time. (It is certainly necessary and the equivalence mapping for the "second stage" duration is trivial.) Townsend (1976b) established this background and then proved some theorems about serial-parallel equivalence and uniqueness. It was actually done with relaxed independence conditions but the assumption of independence permits brevity of exposition here. Note also, that this construction is general to many kinds of processing situations.

With regard to the serial model, let the conditional density on the first processing time when the object is A, be $f_a(t)$ and for B when it is processed first, $f_b(t)$. Then there are two kinds of trials depending on whether A or B happens to be processed first. The probability density on the minimum processing, including the probability of it being A, is just $pf_a(t)$ whereas when the object is B, the density is $(1 - p)\ f_b(t)$. Observe that this order of processing may be under control of the subject, but this does not affect our theoretical results or their implications.

If processing is parallel then the minimum time density when A is done first is $g_a(t)G_b(t)$ where $g_a(t)$ is the actual density on A's processing time and $G_b(t)$ is the probability that B has not been completed by time t. $G_b(t)$ is equivalent to one minus the cumulative distribution function on

B's processing time. It is known in actuarial statistics, quality control and reliability theory as the survivor function. Similarly, B can only be done first at exactly time t if A has not yet been completed so that its density is the mirror image of the other, namely, $g_b(t)G_a(t)$.

Now, if we are to have parallel-serial equivalence then the A-first serial functions must equal the A-first parallel functions for all positive t and similarly for the B-first functions. This leads to the pair of functional equations

$$p\, f_a(t) = g_a(t)\, G_b(t)\,,$$
$$(1-p)\, f_b(t) = g_b(t)\, G_a(t)\,, \qquad (\text{for } t > 0)\,.$$

It was shown that for any well-defined parallel functions, there exists an equivalent serial model, in the sense that p, $f_a(t)$ and $f_b(t)$ exist and are well-defined probability functions of the parallel expressions. On the other hand, there exist serial models, that is, functions $f_a(t)$ and $f_b(t)$ and a p such that no well-defined parallel functions can equal the left-hand side of the above equations (Townsend, 1976b). This means that for experiments that would meet the present conditions, the class of serial models is more general than that of the parallel models. Therefore, no data could satisfy the parallel model without satisfying the serial model as well. However, in principle, there could exist serially produced data that could not be replicated by any parallel models.

Ross and Anderson (1981) developed some statistical tests based on the above results and employed them with some human memory data. We cannot go into details of the experiment here, but the authors were able to conclude that their subjects' data satisfied the conditions for parallelity. That is, parallel processing could not be ruled out in favor of serial processing. On the other hand, because in this case, the serial class encompasses the parallel class, the process could also have been serial. Only if the test had rejected parallelity could a strong conclusion have been drawn in that particular study. Nevertheless, because the authors' theory of memory access predicted parallel processing, the outcome was a propitious one for their theory.

We have developed a number of other experimental designs, based on uniqueness results analogous to the above, so that it may not be necessary to rely wholly on one type of test. In some cases, the parallel models are more general than the serial and in others they are equally general, but each possessing regions of uniqueness that enhances testability (e.g., Townsend and Ashby, 1983; Townsend, 1984; 1989; 1990).

Acknowledgments

The writing of this essay was supported in part by a National Science Foundation – Memory and Cognitive Processes grant No. 8710163. Thanks to Trisha Van Zandt for suggestions and aiding in preparation of the manuscript.

References

1. M. Allais, *Le comportement de l'homme rationnel devant le risque: Critique des postulats et axiomes de l'ecole americaine. Econometrica* **21** (1953) 503–546.
2. J. A. Anderson, *A theory for the recognition of items from short memorized lists, Psychological Re.* **80** (1973) 417–438.
3. J. A. Anderson and G. Hinton, *Models of information processing in the brain* in *Parallel models of associative memory*, eds. G. Hinton and J. A. Anderson (Hillsdale, 1981).
4. J. R. Anderson, *Architecture of cognition* (Harvard University Press, 1983).
5. J. R. Anderson and G. H. Bower, *Human associative memory* (Winston & Sons, 1973).
6. N. H. Anderson, *Foundations of information integration theory* (Academic Press, 1981).
7. M. A. Arbib, *Brains, machines, and mathematics* (McGraw-Hill, 1964).
8. R. C. Atkinson and W. K. Estes, Stimulus sampling theory. In *Handbook of mathematical psychology*, eds. R. D. Luce, R. R. Bush and E. Galanter (Wiley, 1963), vol. III.
9. R. C. Atkinson, J. E. Holmgren and J. F. Juola, Processing time as influenced by the number of elements in a visual display, *Perception & Psychophys.* **6** (1969) 321-326.
10. R. C. Atkinson and R. M. Shiffrin, Human memory: A proposed system and its control processes. In *The psychology of learning and motivation: Advances in research and theory*, eds. K. W. Spence and J. T. Spence (Harper & Row, 1968), vol. 2.
11. F. Attneave, Dimensions of similarity, *Am. J. Psychology* **63** (1950) 516–556.
12. F. Attneave, *Applications of information theory to psychology: A summary of basic concepts, methods, and results* (Holt, Rinehart & Winston, 1959).
13. In *The neurophysiology of Lashley: Selected papers of K. S. Lashley*, eds. F. A. Beach, D. O. Hebb, C. T. Morgan and H. W. Nissen (McGraw-Hill, 1960).
14. E. G. Boring *A history of experimental psychology* (Appleton-Century-Crofts, 1957), 2nd edition.
15. G. H. Bower, Choice-point behavior. In *Studies in mathematical learning theory*, eds. R. R. Bush and W. K. Estes (Standard University Press, 1959).
16. T. H. Brown, P. F. Chapman, E. W. Kairiss and C. L. Keenan, Long-term synaptic potentiation, *Science* **242** (1988) 724–728.

17. B. Burns and B. E. Shepp, Dimensional interactions and the structure of psychological space: The representation of hue, saturation and brightness, *Perception Psychophys.* **43** (1988) 494–507.

18. J. R. Busemeyer, Decision-making under uncertainty: A comparison of simple scalability, fixed-sample, and sequential-sampling models, *J. of Exp. Psychology: Learning, Memory, and Cognition* **11** (1985) 538–564.

19. J. R. Busemeyer, G. M. You and J. T. Townsend, Stochastic approach-avoidance models of decision making. Paper presented at the Twenty-first Annual Mathematical Psychology Society Meeting at Northwestern University, August, 1988.

20. R. R. Bush and F. Mosteller, *Stochastic models for learning* (Wiley, 1955).

21. J. D. Carroll and J. J. Chang, Analysis of individual differences in multidimensional scaling via an *n*-way generalization of "Eckart-Young" decomposition, *Psychometrica* **35** (1970) 283–319.

22. P. Cavanagh, Holographic and trace strength models of rehearsal effects in the item recognition task, *Memory & Cognition* **4** (1976) 186–199.

23. C. H. Coombs, *A theory of data* (Wiley, 1964).

24. F. I. M. Craik and R. S. Lockhart, Levels of processing: A framework for memory research, *J. of Verbal Learning and Verbal Behavior* **11** (1972) 671–684.

25. R. DeValois and K. DeValois, Neural coding of color. In *Handbook of perception*, eds. E. C. Carterette and M. P. Friedman (Academic Press, 1975), vol. 5.

26. W. Edwards, Probability-preferences in gambling, *Am. J. Psychology* **66** (1953) 349–364.

27. W. Edwards, Theory of decision making, *Psychological Bull.* (1954) 380–417.

28. W. Edwards, Behavioral decision theory. In *Ann. Rev. Psychology*, eds. P. R. Farnsworth, O. McNema and Q. McNema (Annual Review Inc, 1961) 473–498.

29. W. Edwards, Subjective probabilities inferred from decisions, *Psychological Rev.* **69** (1962) 109–135.

30. H. Egeth, Parallel vs serial processes in multidimensional stimulus discrimination, *Perception & Psychophys.* **1** (1966) 245–252.

31. H. Egeth, J. Jonides and S. Wall, Parallel processing of multi-element displays, *Cognitive Psychology* **3** (1972) 674–698.

32. J. M. Eich, A composite holographic associative recall model, *Psychological Rev.* **89** (1982) 627–661.

33. H. Eisler and E. E. Roskam, Multidimensional similarity: An experimental and theoretical comparison of vector, distance, and set theoretical models. II. Multidimensional analyses: The subjective space, *Acta Psychologica* **41** (1977) 335–363.

34. D. Ellsberg, Risk, ambiguity, and the Savage axioms, *Quarterly J. Econ.* **75** (1961) 643–669.

35. S. Embretson, Multicomponent latent trait models for test design. In *Test design: Developments in psychology and psychometrics*, eds. S. Embretson (Academic Press, 1985).

36. W. K. Estes, Toward a statistical theory of learning, *Psychological Rev.* **57** (1950) 94–107.

37. W. K. Estes, The statistical approach to learning theory. In *Psychology: A study of a science*, ed. S. Koch (McGraw-Hill, 1959), vol. 2.

38. W. K. Estes, Learning theory, *Ann. Rev. Psychology* **13** (1962) 107–144.

39. W. K. Estes, The cognitive side of probability learning, *Psychological Rev.* **83** (1976) 37–64.

40. P. C. Fishburn, Binary choice probabilities between gambles: Interlocking expected utility, *J. Math. Psychology* **14** (1976) 99–122.

41. D. L. Fisher and W. M. Goldstein, Stochastic PERT networks as models of cognition: Derivation of the mean, variance, and distribution of reaction time using order-of-processing (OP) diagrams, *J. Math. Psychology* **27** (1988) 121–151.

42. K. Fukushima, A neural network model for selective attention in visual pattern recognition, *Biological Cybernetics* **55** (1986) 5–15.

43. Galton, *Inquiries into human faculty and its development* (Dent, 1883).

44. W. M. Goldstein and H. J. Einhorn, Expression theory and the preference reversal phenomena, *Psychological Rev.* **94**(2), (1987) 236–254.

45. D. M. Green and J. A. Swets, *Signal detection theory and psychophysics* (Krieger, 1966).

46. S. Grossberg, How does the brain build a cognitive code, *Psychological Rev.* **87** (1980) 1–51.

47. In *Neural networks and natural intelligence*, ed. S. Grossberg (MIT Press, 1988).

48. E. Hearst, Aversive conditioning and external stimulus control. In *Punishment and aversive behavior*, eds. B. Campbell and R. N. Church (Appleton-Century-Crofts, 1969).

49. D. O. Hebb, *Organization of behavior* (Wiley, 1949).

50. D. L. Hintzman, Schema abstraction in multiple-trace memory model, *Psychological Rev.* **93** (1986) 429–445.

51. J. J. Hopfield and D. W. Tank, Computing with neural circuits: A model, *Science* **233** (1986) 625–633.

52. C. L. Hull, *Principles of behavior, an introduction to behavior theory* (Appleton-Century-Crofts., 1943).

53. C. L. Hull, *A behavior system: an introduction to behavior theory concerning the individual organism* (Yale University Press, 1952).

54. M. S. Humphreys, J. D. Bain and R. Pike, Different ways to cue a coherent memory system: A theory for episodic, semantic and procedural tasks, *Psychological Rev.* **96** (1989) 208–233.

55. E. B. Hunt, *Artificial intelligence* (Academic Press, 1975).

56. L. M. Hurvich and D. Jameson, An opponent-process theory of color vision, *Psychological Rev.* **64** (1957) 384–404.

57. D. Kahneman and A. Tversky, Prospect theory: An analysis of decision under risk, *Econometrics* **47** (1979) 263–291.

58. C. Koch, T. Poggio and V. Torre, Retinal ganglion cells: A functional interpretation of dendritic morphology, *Philosophical Transactions of the Roy. Soc. London* **B298** (1982) 227–264.

59. T. Kohonen, *Associative memory: A system-theoretic approach* (Springer-Verlag, 1977).

60. D. H. Krantz, Rational distance function for multidimensional scaling, *J. Math. Psychology* **4** (1967) 226–245.

61. D. H. Krantz, Measurement theory and qualitative laws in psychophysics, In *Contemporary developments in mathematical psychology*, eds. D. H. Krantz, R. C. Atkinson, R. D. Luce and P. Suppes (Freeman, 1974), vol. 2.

62. D. H. Krantz, R. D. Luce, P. Suppes and A. Tversky *Foundations of Measurement*, (Academic Press, 1971), vol. 1.

63. L. E. Krueger, A theory of perceptual matching, *Psychological Rev.* **85** (1978) 278–304.

64. J. B. Kruskal, Multidimensional scaling by optimizing goodness of fit to a non-metric hypothesis, *Psychometrica* **29** (1964a) 1–28.

65. J. B. Kruskal, Nonmetric multidimensional scaling: A numerical method, *Psychometrica* **29** (1964b) 115–130.

66. K. S. Lashley, *Brain mechanisms in intelligence* (University of Chicago Press, 1929).

67. K. S. Lashley, Integrative functions of the cerebral cortex. *Physiological Rev.* **13** (1933) 1–42.

68. V. A. Lefebvre, *Algebra of conscience* (Reidel, 1982).

69. V. A. Lefebvre, The fundamental structure of human reflexion, *J. Social Biological Structure* **10** (1987) 129–175.

70. D. S. Levine, Neural population modeling and psychology: A review, *Math. Biosci.* **66** (1983) 1–86.

71. K. Lewin, *A dynamic theory of personality* (McGraw-Hill, 1935).

72. K. Lewin, *Principles of topological psychology* (McGraw-Hill, 1936).

73. S. W. Link and R. A. Heath, A sequential theory of psychological discrimination, *Psychometrika* **40** (1975) 77–105.

74. M. Livingstone and D. Hubel, Segregation of form, color, movement and depth: Anatomy, physiology and perception, *Science* **240** (1988) 740–749.

75. R. D. Luce, *Individual choice behavior* (Wiley, 1959).

76. R. D. Luce, *Response times* (Oxford University Press, 1986).

77. R. D. Luce and D. M. Green, A neural timing theory for response times and psychophysics of intensity, *Psychological Rev.* **79** (1972) 14–57.

78. R. D. Luce and L. Narens, The mathematics underlying measurement on the continuum, *Science* **236** (4808) (1987) 1527–1532.

79. D. Marr, *Vision* (Freeman, 1982).

80. D. W. Massaro, Stimulus information vs processing time in auditory pattern recognition, *Perception & Psychophys.* **12** (1972) 50–56.

81. D. W. Massaro, The computer as a metaphor for psychological inquiry: Considerations and recommendations, *Behavior Research methods. Instruments & Computers* **18** (1986) 73–92.

82. *Parallel distributed processing: Exploration in the microstructure of cognition, vol. 2 Psychological and biological models*, eds. J. L. McClelland and D. E. Rumelhart (MIT Press, 1986).

83. W. S. McCulloch and W. H. Pitts, A logical calculus of the ideas immanent in nervous activity, *Bull. Math. Biophys.* **5** (1943) 115–133.

84. W. J. McGill, Stochastic latency mechanisms. In *Handbook of mathematical psychology*, eds. R. D. Luce, R. R. Bush and E. Galanter (Wiley, 1963), vol. 1.

85. N. E. Miller, Liberalization of basic S-R concepts: Extensions to conflict behavior, motivation, and social learning. In *Psychology: A study of a science*, ed. S. Koch (McGraw-Hill, 1959), vol. 2.

86. M. Minsky and S. Pappert, *Perceptrons: An introduction to computational geometry* (MIT Press, 1969).

87. J. M. Miyamoto, Generic utility theory: Measurement foundations and applications in multiattribute utility theory, *J. Math. Psychology* **32** (1988) 357–404.

88. J. M. Miyamoto and S. A. Eraker, A multiplicative model of the utility of survival duration and health quality, *J. Experimental Psychology: General* **117** (1988) 3–20.

89. B. B. Murdock, Jr., A parallel-processing model for scanning, *Perception & Psychophys.* **10** (413) (1971) 289–291.

90. B. B. Murdock, Jr., A theory for the storage and retrieval of item and associative information, *Psychological Rev.* **89** (1982) 609–626.

91. M. Nakagawa, A mathematical model of approach and avoidance behavior in the psychological field, *Jpn. Psychological Res.* **29** (1987) 59–70.

92. L. Narens, *Abstract measurement theory* (MIT Press, 1985).

93. A. Newell, Symbolic processing and human cognitive architecture. Paper presented at the Electrical Engineering Distinguished Lecture Series, Purdue University, West Lafayette, IN, Fall, 1988.

94. A. Newell and H. A. Simon, Computers in psychology. In *Handbook of mathematical psychology*, eds. R. D. Luce, R. R. Bush and E. Galanter (Wiley, 1963), vol. 1.

95. D. A. Norman and D. E. Rumelhart, *Exploration in cognition* (Freeman, 1975).

96. R. G. Pachella, The interpretation of reaction time in information processing research. In *Human information processing: Tutorials in performance and cognition*, ed. B. H. Kantowitz (Erlbaum, 1974).

97. R. G. Pachella, J. E. K. Smith and K. E. Stanovich, Qualitative error analysis and speeded classification. In *Cognitive theory*, eds. N. J. Castellan, Jr. and F. Restle (Lawrence Erlbaum Associates, 1978), vol. 3.

98. J. W. Payne, J. R. Bettman and E. J. Johnson, Adaptive strategy selection in decision making, *J. Experimental Psychology: Learning, Memory, & Cognition* **14**(3) (1988) 534–552.

99. W. W. Peterson and T. G. Birdsall, The theory of signal detectability. Electronic Defense Group, University of Michigan, Technical Report No. 13, Sept., 1953.

100. J. Pfanzagl, *Theory of measurement* (Wiley, 1968).

101. R. Pike, Comparison of convolution and matrix distributed memory systems for associative recall and recognition, *Psychological Rev.* **91** (1984) 281–294.

102. T. Poggio, V. Torre and C. Koch, Computational vision and regularization theory, *Nature* **317** (1985) 314–319.

103. M. T. Posner, *Cognition: An introduction* (Scott, Foresman & Co, 1973).

104. K. H. Pribram, M. Nuwer and R. J. Baron, The holographic hypothesis of memory structure in brain function and perception. In *Contemporary developments in mathematical psychology*, eds. D. H. Krantz, R. C. Atkinson, R. D. Luce and P. Suppes (Freeman, 1974), vol. 2.

105. J. G. W. Raaijmakers and R. M. Shiffrin, Search of associative memory, *Psychological Rev.* **88** (1981) 93–134.

106. R. Ratcliff, A theory of memory retrieval, *Psychological Rev.* **85** (1978) 59–108.

107. R. Ratcliff, Parallel-processing mechanisms and processing of organized information in human memory. In *Parallel models of associative memory*, eds. G. Hinton and J. A. Anderson (Erlbaum, 1981).

108. F. Restle, *Psychology of judgment and choice* (Wiley, 1961).

109. D. M. Riefer and W. H. Batchelder, Multinomial modeling and the measurement of cognitive processes, *Psychological Rev.* **95** (3) (1988) 318–339.

110. F. S. Roberts, *Measurement theory with applications to decision making, utility, and the social sciences* (Addison-Wesley, 1979).

111. H. L. Roediger, Memory metaphors in cognitive psychology, *Memory and Cognition* **8**(3) (1980) 231–246.

112. F. Rosenblat, *Principles of neurodynamics* (Spartan Books, 1962).

113. B. H. Ross and J. R. Anderson, A test of parallel versus serial processing applied to memory retrieval, *J. Math. Psychology* **24** (1981) 183–223.

114. In *Parallel distributed processing: Explorations in the microstructure of cognition. vol. 1: Foundations*, eds. D. E. Rumelhart and J. L. McClelland (MIT Press, 1986).

115. L. J. Savage, *The foundations of statistics* (Wiley, 1954).

116. W. Schneider, Connectionism: Is it a paradigm shift for psychology? *Behavioral Research Methods, Instruments, & Computers* **19** (1987) 73–83.

117. W. Schneider and R. M. Shiffrin, Controlled and automatic human information processing: I. Detection, search, and attention. *Psychological Rev.* **84** (1977) 1–66.

118. P. H. Schönemann, Factorial definitions of intelligence: Dubious legacy of dogma in data analysis. In *Multidimensional data representations: When & Why*, ed. I. Borg (Mathesis Press, 1981).

119. P. H. Schönemann, Some algebraic relations between involutions, convolutions, and correlations, with applications to holographic memories. *Biological Cybernetics* **56** (1987) 367–374.

120. R. Schweickert, A critical path generalization of the additive factor method: Analysis of a Stroop task, *J. Math. Psychology* **18** (1978) 105–139.

121. R. Schweickert, Latent network theory: Scheduling of processes in sentence verification and the Stroop effect. *J. Experimental Psychology: Learning, Memory, and Cognition* **9** (1983) 353–383; R. Schweickert and J. T. Townsend, A trichotomy method: Interactions of factors prolonging sequential and concurrent mental processes in stochastic (PERT) networks, *J. Math. Psychology* **33** (1989) 328–347.

122. A. C. Scott, Neurodynamics: A critical survey. *J. Math. Psychology* **15** (1977) 1–45.

123. C. E. Shannon, A mathematical theory of communication. *Bell Systems Technical J.* **27** (1948) 379–423.

124. C. E. Shannon and W. Weaver, *The mathematical theory of communication* (University of Illinois Press, 1949).

125. R. N. Shepard, Attention and the metric structure of the stimulus space, *J. Math. Psychology* **1** (1964) 54–87.

126. B. F. Skinner, Methods and theories in the experimental analysis of behavior, *Behavioral and Brain Sci.* **7** (1984) 511–546.

127. E. E. Smith, Choice reaction time: An analysis of the major theoretical positions, *Psychological Bull.* **69** (1968) 77–110.

128. J. G. Snodgrass, Reaction times for comparisons of successively presented visual patterns: Evidence for serial self-terminating search, *Perception & Psychophys.* **12** (1972) 364–372.

129. J. G. Snodgrass and J. T. Townsend, Comparing parallel and serial models: Theory and implementation, *J. Exp. Psychology: Human Perception and Performance* **6**(2) (1983) 330–354.

130. C. Spearman, General intelligence, objectively determined and measured, *Am. J. Psychology* **15** (1904) 201–293.

131. C. Spearman, *The abilities of man* (Macmillan Co, 1927).

132. R. Sternberg, Toward a triarchic theory of human intelligence, *Behavior and Brain Sci.* **7** (1984) 269–315.

133. S. Sternberg, Stochastic learning theory, In *Handbook of mathematical psychology*, eds. R. D. Luce, R. R. Bush and E. Galanter (Wiley, 1963), vol. 2.

134. S. Sternberg, High-speed scanning in human memory, *Science* **153** (1966) 652–654.

135. S. Sternberg, The discovery of processing stages: Extensions of Donder's method. In *Attention and performance*, eds. W. G. Koster (North-Holland Press, 1969), vol. 2.

136. S. Sternberg and R. L. Knoll, The perception of temporal order: Fundamental issues and a general model. In *Attention and performance*, ed. S. Kornblum (Academic Press, 1973), vol. 4.

137. S. S. Stevens, On the theory of scales of measurement, *Science* **103** (1946) 677–680.

138. S. S. Stevens, Mathematics, measurement and psychophysics. In *Handbook of experimental psychology*, ed. S. S. Stevens (Wiley, 1951); P. Suppes and J. L. Zinnes, Basic measurement theory. In *Handbook of Mathematical Psychology*, eds. R. D. Luce, R. R. Bush and E. Galanter (Wiley, 1963), vol. 1.

139. W. P. Tanner and J. A. Swets, A decision-making theory of visual detection, *Psychological Rev.* **61** (1954) 401–409.

140. J. Theios, Reaction time measurements in the study of memory processes: Theory and data. In *The Psychology of learning and motivation*, ed. G. H. Bowen (Academic Press, 1973), vol. 7.

141. J. Theios, P. G. Smith, S. E. Haviland, J. Traupmann and M. C. Moy, Memory scanning as a serial self-terminating process, *J. Exp. Psychology* **97** (1973) 323–336.

142. L. L. Thurstone, *Multiple-factor analysis* (University of Chicago Press, 1947).

143. E. C. Tolman, Principles of purposive behavior. In *Psychology: A study of a science*, ed. S. Koch (McGraw-Hill, 1959), vol. 2.

144. W. S. Torgerson, *Theory and methods of scaling* (Wiley, 1958).

145. J. T. Townsend, Mock parallel and serial models and experimental detection of these. *Purdue Centennial symposium on Information Processing* (Purdue University Press, 1969).

146. J. T. Townsend, A note on the identifiability of parallel and serial processes. *Perception & Psychophys.* **10** (1971a) 161–163.

147. J. T. Townsend, Alphabetic confusion: A test of models for individuals, *Perception & Psychophys.* **9** (1971b) 449–454.

148. J. T. Townsend, Some results concerning the identifiability of parallel and serial processes. *Br. J. Math. Stat. Psychology* **25** (1972) 168–199.

149. J. T. Townsend, Issues and models concerning the processing of a finite number of inputs. In *Human information processing: Tutorials in performance and cognition* ed. B. H. Kantowitz Hillsdale (Erlbrum, 1974).

150. J. T. Townsend, A stochastic theory of matching processes, *J. Math. Psychology* **14** (1976a) 1–52.

151. J. T. Townsend, Serial and within-stage independent parallel model equivalence on the minimum completion time, *J. Math. Psychology* **14** (1976b) 219–238.

152. J. T. Townsend, Some characteristics of visual whole report behavior, *Acta Psychologica* **47** (1981) 149–173.

153. J. T. Townsend, Book review of *Algebra of Conscience* by Vladimir A. Lefebvre, *J. Math. Psychology* **27** (1983) 461–471.

154. J. T. Townsend, Uncovering mental processes with factorial experiments, *J. Math. Psychology* **28** (1984) 363–400.

155. J. T. Townsend, Serial vs Parallel processing: Sometimes they look like Tweedledum and Tweedledee but they can (and should) be distinguished. *Psychological Science* **1** (1990) 46–54.

156. J. T. Townsend and F. G. Ashby, Methods of modeling capacity in simple processing systems. In *Cognitive Theory*, eds. J. Castellan and F. Restle (Erlbaum, 1978), vol. 3, 200–239.

157. J. T. Townsend and F. G. Ashby, *Stochastic modeling of elementary psychological processes* (Cambridge University Press, 1983).

158. J. T. Townsend and J. R. Busemeyer, Approach-avoidance: Return to dynamic decision behavior. In *Current issues in cognitive processes: The Tulane Floweree Symposium on cognition*, ed. C. Izawa (Erlbaum) 1989; J. T. Townsend and R. Schweickert, Toward the trichotomy method: Laying the foundation of stochastic mental networks, *J. Math. Psychology* **33** (1989) 309–327.

159. A. Tversky, Utility theory and additivity analysis of risky choices, *J. Exp. Psychology* **75** (1967) 27–37.

160. A. Tversky, The intransitivity of preferences, *Psychological Rev.* **76** (1969) 31–48.

161. A. Tversky, Elimination by aspects, *Psychological Rev.* **79** (1972a) 281–299.

162. A. Tversky, Choice by elimination, *J. Math. Psychology* **9** (1972b) 341–367.

163. A. Tversky and D. Kahneman, The framing of decisions and the psychology of choice, *Science* **211** (1981) 453–458.

164. J. von Neumann, *The computer and the brain* (Yale University Press, 1958).

165. J. von Neumann and O. Morgenstern, *Theory of games and economic behavior* (Princeton University Press, 1953).

166. D. Vorberg, On the equivalence of parallel and serial models of information processing. Paper presented at the Tenth Annual Mathematical Psychology Meetings held at University of California at Los Angeles, 1977).

167. W. A. Wickelgren, Multitrace strength theory. In *Models of human memory*, ed. D. A. Norman (Academic Press, 1970).

168. D. J. Willshaw, Models of distributed associative memory (University of Edinburgh, 1971).

169. D. J. Willshaw, In *Parallel models of associative memory*, eds. G. Hinton and J. A. Anderson (Erlbaum, 1981).

170. J. I. Yellott, Correction for fast guessing and the speed-accuracy trade-off in choice reaction time. *J. Math. Psychology* 8 (1971) 159–199.

Ariadne's Thread: The Role of Mathematics in Physics

Hans C. von Baeyer
Department of Physics
College of William and Mary
Williamsburg, VA 23185, USA

1. Reading Eugene Wigner's essay[1] on the role of mathematics in science, the physics teacher is struck by the sentence: "(Newton's law of planetary motion,) particularly since a second derivative appears in it, is simple only to the mathematician, not to common sense or to non-mathematically-minded freshmen." The statement is undoubtedly true, but why? Why do freshmen seem to have a harder time with physics than with French, with political science, or, for that matter, with mathematics? What is so difficult about physics? Thinking about this question leads to three observations that bear on the more general problem set forth by Wigner.

The emphasis on the second derivative puts the blame for the difficulty of physics on mathematics, but that is only part of the whole story. To be sure, when asked about the source of their difficulties with physics, freshman almost invariably answer: "It's the math. I understand the concepts, but I can't do the math." Indeed, correcting freshman physics exams suggests that, on a superficial level, inadequate proficiency in mathematics is the culprit. Even the most elementary conventions of mathematical notation are broken, and the simplest operations are botched. But experience indicates that mathematical errors are a symptom, not the cause, of the difficulty. The same students who fail physics exams frequently do well in calculus. As is often the case, they have it backward. They can do the math, but they do not understand the concepts.

What seems to happen is that they temporarily forget what they have learned previously. This loss of mathematical skills as applied to difficult concepts has its counterpart in verbal expression. Teachers of writing have noticed that when students are confronted with complex or confusing ideas, the first thing that breaks down is their basic writing skill. As they struggle to express thoughts that are not yet fully articulated, words and sentences get muddled up, punctuation fails, and eventually even spelling suffers. The same student who can write a witty and lucid essay about her cat is reduced to near illiteracy in her analysis of the concept of justice in "King Lear". Her problem is not with writing *per se*, but with the challenge of doing two things at once: thinking and writing. In a similar way my French, which is normally adequate, degenerates into gibberish when I try to communicate complicated ideas, even if I know the appropriate technical vocabulary. So what seems to be a deficiency in mathematics is often a symptom, not a root cause, of the difficulty of physics.

If Wigner's statement were generalized beyond the elementary example of Newton's law of planetary motion, it would not be true. Physics, in general, is not simple for the "mathematically-minded" freshman, as Wigner calls him, or even the mathematician. If it were, all mathematicians would be good at physics – but they are not. The difficulty of physics must be sought at a deeper level.

Unwittingly, the freshman who protests, erroneously, that he understands the concepts, but cannot do the math, puts his finger on the crucial distinction. The difficulty of physics is *conceptual*, rather than *mathematical*. Even the best students have serious conceptual difficulties in introductory physics classes,[2] difficulties which carry over into advanced courses. Persistent misconceptions can be found in all branches of physics, but for the sake of illustration, it is sufficient to consider mechanics, and in particular the concepts of motion, velocity, and acceleration.

It is important to distinguish between the uses of these three concepts in physics and in mathematics. The first, motion, though meaningful in physics, is not particularly useful in mathematics. Velocity and acceleration, on the other hand, are too easily thought of as derivatives, and therefore classified, as Wigner implied, among mathematical concepts. They are not. They refer to real objects, and it is precisely in the linking between their crisp mathematical definitions and the concrete object they refer to (a car, for example), that the trouble arises. It is not acceleration as second derivative that puzzles students, but acceleration as a property or attribute of an object.

The difficulty of concepts in physics increases with their degree of abstraction. (In the past, the terms "concept" and "abstraction" were used almost interchangeably in the psychological literature,[3] to signify an idea as opposed to a thing, but since the former is now preferred, the latter can be employed for a more specific purpose.) The word "abstract" derives from the Latin, meaning to pull away, and is used here to signify removal from concrete objects. In that sense motion, velocity, and acceleration are concepts of increasing degree of abstraction. When a non-scientist sees a car, he sees bumpers, wheels, and windows. He can tell whether the car is moving or not, and he has a qualitative sense of its speed and direction. A physicist, on the other hand, endows that same car with invisible velocity and acceleration vectors. This association between objects and abstract concepts does not, as Wigner correctly implies, follow from common sense. For this reason freshmen, when given a verbal description, or even a visual demonstration, of a specific example of motion, have difficulty abstracting the velocity, and especially the acceleration. Conversely, given the mathematical description of the motion, they find it difficult to imagine the corresponding behavior of a real object. The abstraction, the gulf between the thing and the formula, is too great.

It is not abstractness by itself that causes difficulty. Mathematics deals almost exclusively with abstract concepts, and yet is easier than physics for many students. The intellectual hurdle seems to be the need to shuttle back and forth between the real world and its abstract description, between the car and its acceleration. That this shuttling may be the cause of the breakdown of elementary mathematics skills among freshman can be seen in examination papers. Even where the equations are perfectly correct, physical assumptions and arguments interrupt and ruin their straightforward mathematical solution. Conceptual considerations interfere with mathematical manipulations. At the professional level, the relationship between the objective world and the mathematical model also causes trouble. This relationship is the locus of the theory of measurement, and, more broadly, of the thorny philosophical questions of interpretation that most physicists sweep under the rug. The process of abstraction, then, emerges as one of the sources of the difficulty of physics at all levels.

The first lesson we learn from the difficulties experienced by freshman is that physics has two distinct, though intertwined, components – the mathematical and the conceptual. Concepts, in turn, vary in degree of abstraction. In this light it appears that the metaphor of mathematics as the language of physics, which pervades the entire discussion of the effectiveness of mathematics, is misleading. A language is a symbolic system sufficient to describe its universe of discourse: In English, I can express any idea without recourse to any other language. Physics, on the other hand, needs not only mathematics, but, in addition, a rich fabric of interrelated concepts. Mathematics, by itself, is not sufficient. In fact, the most profound issues in physics are not mathematical but conceptual. By subtly overstating the importance of mathematics, the "language" metaphor fails to capture the significance of the conceptual content of physics, and should therefore be used with caution.

An alternative metaphor might liken the concepts of physics to the organs of the body. They are related to each other, sometimes even overlapping, they depend on each other, and a few are not even precisely defined. Mathematics, in this image, is represented by the nervous system that connects the parts in an orderly, logical fashion, and assures adherence to such general principles as causality. The network of mathematics organizes and animates the whole body of physics, but it does not stand alone. The system of concepts is required to flesh it out.

2. Close reading of Wigner's remark about freshman suggests a second observation, besides the distinction between mathematical and conceptual difficulties. The physics teacher is struck by the enormous variety and scope of the misconceptions he encounters. Students reproduce every possible permutation of the primitive conceptual ingredients of the notions of velocity and acceleration. In addition, they invent highly original and outrageous-sounding definitions and arguments of their own. Rather than dismissing these aberrations as irrelevant to the pursuit of physics, one can learn from them.

Variety is an indication of *complexity*. In a simple device, such as a switch, there is only one choice – on or off – and hence, only one type of error. Complex systems, on the other hand, allow a greater variety of choices and hence, of errors. In this light, it appears that the concept of acceleration, for example, is even more complex than Wigner suggests when he calls the second derivative "not a very immediate concept." The complexity of the concept of acceleration does not become fully apparent in correct applications, but is adumbrated by the variety of errors made by freshman. Examples of such errors range from simple linguistic boners, such as the notion that since weight is mass times the acceleration of gravity, a non-accelerating object has no weight, to the subtle distinction, in three dimensions, between acceleration and the rate of change of speed. Multiplied a hundredfold, such errors hint at the complexity of the concept of acceleration.

Indeed, when examined closely, both the mathematical and physical definitions of acceleration turn out to be complex. Regarded as the derivative of a vector, the definition presupposes the cumbersome machinery of differential calculus, as well as the formalism of vectors. If, as Wigner suggests, the second derivative is defined in terms of the reciprocal of the radius of the osculating circle to the curve of position versus time, graphical complications take the place of the complexities of calculus. Regarded, on the other hand, as a measurable attribute of a moving object, acceleration is just as complex. Even in the case of one-dimensional motion, the operational definition of acceleration requires at least three measurements of position at specified times, several steps of arithmetic to extract the average acceleration, and finally the intellectually formidable task of extrapolation to the limit of instantaneous acceleration. All this, as the car is speeding by!

Complexity, in any case, is relative. It depends on the selection of the fundamental axioms of the system. Thus, for example, the equation $1 + 1 = 2$ is simple, in the context of arithmetic, but in the *Principia Mathematica*, it emerges only after more than 700 pages of closely reasoned argument[4] because in the context of logic it is complex. Similarly, a resistor is a simple one-parameter device in terms of electrical engineering, but requires 10^{23} parameters in terms of solid state physics. Freshmen instinctively dip below the superficial simplicity of the concept of acceleration and reveal, by their mistakes, its hidden complexity.

There is another way in which this complexity can be demonstrated. The history of the concept of motion is long and convoluted. The interval between the discovery of the paradoxes of motion by Zeno, and their resolution by Newton, was twenty-one centuries. The significance of that long struggle was summed up by the historian Herbert Butterfield, quoted by Gerald Holton[5]: "Of all the intellectual hurdles which the human mind has been faced with and has overcome ... the one which seems to me to have been the most amazing in character and the most stupendous in the scope of its consequences is the one relating to the problem of motion." Only with the arrogance born of hind-sight can we call the analysis of motion in any sense "simple".

These considerations make the confusion of freshmen a little more understandable. In two and a half millennia of Western science, every conceivable misconception about motion, velocity, and acceleration has cropped up, and occasionally even been published. If ontogeny recapitulates philogeny, if the development of a concept in an individual mind follows the course of its historical development, then the errors that students make during their lightning trip through mechanics are merely the echoes of all the errors made by the great thinkers of the past.

What is the role played by mathematics in this view? When a freshman text develops Newton's law of planetary motion, it follows a clear, logical path from beginning to end, avoiding all the misconceptions and conceptual errors that misled thinkers in antiquity and mislead students today. What guides the exposition is formalized logic, expressed in the symbolic notation of mathematics. To make this arrangement possible, concepts must have quantifiable components, handles, as it were, that allow connections to be made with other concepts via mathematics. The concept of acceleration, for example, which, as we have seen, is more than just mathematical, does incorporate the second derivative, and this, in turn, allows connections to be made to other quantified concepts, such as mass and force.

In the exposition of the law of planetary motion, mathematics is the nerve that leads infallibly from concept to concept until the whole structure emerges. If one remembers not only the useful concepts, however, but also the countless misconceptions and errors that the rigorous mathematical development avoids without touching, then mathematics begins to resemble not as much a nerve as the thread that Ariadne used to guide her lover Theseus out of the Labyrinth in which he slew the dreaded Minotaur. The Labyrinth, in that case, stands for the immensely complex system of useful concepts and misconceptions, false trails and blind alleys that make up the body of physics.

3. Conceptual complexity and the proliferation of dead ends suggest a model for the origin of physics. Richard Dawkins, updating Darwin, has demonstrated[6] how the marvellously complex design we see in the living world around us could have arisen, without the intervention of a Designer, through evolution. The necessary prerequisites for this process are a mechanism for *cumulative selection* and a *selection principle.* Cumulative selection differs from one-step selection in that each random step is not an entirely new deal, like a coin toss, but follows from, and builds on, the previous step. In nature, accumulation of small evolutionary steps is assured by the laws of heredity. The selection principle that guides and shapes living systems and thereby provides what seems to be a short term "goal" for the evolutionary process, is survival, or, equivalently, reproductive success. In the development of physics, accumulation of changes is guaranteed by the collective memory of civilized society, as formalized by written publication of results. The selection principle that allows some changes to survive while discarding others as dead ends is the requirement that theories be both mathematically consistent and experimentally verifiable. With the two prerequisites for evolution in place, it is tempting to try to describe the historical development of the marvellous complexity of physics in evolutionary terms.

Evolution has been invoked before in this connection. The last of the four partial solutions to Wigner's problem listed by the mathematician Richard Hamming[7] is entitled "The evolution of man provided the model."

This solution is not as simple and convincing as the preceding three but if I understand it correctly, it involves the evolution of the human brain towards the abilities to create both mathematics and useful models of reality and thus, natural science. Hamming estimates that there have only been at most two hundred generations of scientific thinkers, in the modern sense of the word – far too few to allow evolution to make significant changes. He therefore concludes that evolution cannot contribute very much to the solution of the problem of the effectiveness of mathematics. What I suggest here is altogether different. It concerns not the evolution of the brain, but the evolution of physics, i.e., of a branch of knowledge. The "generations" in this case are not human generations but generations of concepts. Since productive scientists invent several concepts a week, or, if they are in the genius class, several per hour, the number of individual concepts and hence, of generations of concepts, is many orders of magnitude greater than two hundred, making evolution at least conceivable.

Even with that distinction, any attempt to apply the theory of evolution to systems outside the realm of biology, is a risky undertaking. In his eloquent critique of sociobiology[8] the French biologist François Jacob warns that: "... so great appears to be the need of the human mind for unity and coherence of explanation that any theory of some importance is liable to be overused and slip into myth." He believes that this has happened to the theory of evolution and that it has therefore been adduced in contexts where it has no business appearing. Specifically, Jacob writes: "There is no more reason to seek an evolutionary explanation for moral codes than there would be to seek such an explanation for poetry or physics." Since he does not elaborate on this remark, I must assume that Jacob, like Hamming, refers to biological evolution, not the evolution of ideas. Nevertheless, the warning is apposite. It is easy to suggest that evolution explains the origin of complex systems, but difficult to prove it.

An evolutionary theory of the development of physics is beyond the scope of this essay. The background for such an attempt has been laid out in a more general context by the philosopher Stephen Toulmin in a comprehensive analysis of the nature of rationality. Using Darwin's theory as model, Toulmin examines the way concepts are formed and used and how they evolve with time. In sharp disagreement with the revolutionary theory of Thomas Kuhn, Toulmin[9] writes about the concept of motion: "The 'physical' problems of the 1330s – the scholastic analysis of the terms available for characterizing different varieties of 'change' or 'motion' – have been transformed bit by bit, through six centuries of intellectual development, into the contemporary problems of relativity theory and quantum mechanics. At no point, however, has this process been rationally discontinuous."

The question of the unreasonable effectiveness of mathematics can now be asked afresh, from an evolutionary point of view: "How did the unreasonable effectiveness of mathematics in the natural sciences develop?" With the analogy between mathematics and the body's nervous system in mind, this question is reminiscent of the perennial question: "How could the human eye evolve, through random mutations, into such an incredibly complex and powerful organ?" Biology is well on its way to answering the latter question. Perhaps physicists can learn something from their colleagues in the life sciences to help them answer the former. Following standard procedure, they should take into account *all* the data, and that includes, in addition to the spectacular successes of physics, its innumerable false starts and dead ends throughout history, its fossil record, as it were. Perhaps an evolutionary perspective on Wigner's question will deprive it of a little of its mystery.

References

1. E. P. Wigner, *The Unreasonable Effectiveness of Mathematics in the Natural Sciences, Commun. Pure Appl. Math.* **13** (1960) 1–14; reprinted in *The World of Physics*, ed. J. H. Weaver (Simon and Schuster, 1987), vol. 3.
2. C. P. Peters, *Even Honors Students Have Conceptual Difficulties With Physics, Am. J. Phys.* **50** (1982) 501–508.
3. A. Pikas, *Abstraction and Concept Formation* (Harvard University Press, 1966), p. 228.
4. E. Nagel, *Symbolic Notation, Haddock's Eyes and the Dog-Walking Ordinance*, in *The World of Mathematics*, ed. J. R. Newman (Simon and Schuster, 1956), vol. 3, p. 1894.
5. G. Holton, *The Changing Allegory of Motion*, in *Thematic Origins of Scientific Thought* (Harvard University Press, 1973).
6. R. Dawkins, *The Blind Watchmaker* (W. W. Norton & Co., 1986).
7. R. W. Hamming, *The Unreasonable Effectiveness of Mathematics, Am. Math. Monthly* **87** (1980) 81–90; reprinted in *Mathematical Analysis of Physical Systems*, ed. R. E. Mickens (Van Nostrand Reinhold Co., 1985).
8. F. Jacob, *Myth and Science*, in *The Possible and the Actual* (Pantheon Books, 1982), pp. 22–24.
9. S. Toulmin, *Human Understanding* (Princeton University Press, 1972), vol. I, p. 253.

The Disproportionate Response

Bruce J. West
Department of Physics
University of North Texas
Denton, TX 76203, USA

1. Introduction

In a now classic paper Wigner[1] called attention to the unreasonable effectiveness of mathematics in the natural sciences. He made two major points in his discussion:

1) mathematical concepts turn up in entirely unexpected connections

and

2) we do not know if a particular mathematical representation of a phenomenon is unique because we do not understand the reasons for its usefulness.

Two decades later, Hamming[2] wrote a rejoinder in which he sketched his reasons why he believed this is true:

1) we see what we look for,
2) we select the kind of mathematics to use,
3) science in fact answers comparatively few problems; and
4) the evolution of man provided the model.

In the end, Hamming agrees with Wigner and is not convinced by his own arguments. He does not feel that his four reasons are sufficient to explain the logical (mathematical) side of nature. Herein, we explore a number of distinct but related questions having to do with the interrelations between scientific facts and mathematical models.

The image of the scientist in grand isolation sifting through vast collections of facts; organizing, building and ultimately explaining previously mysterious phenomena is poignantly poetic but actually has little to do with the working scientist. The image must also include the realization that a scientist in this age or any age has a continuity of vision, purpose and interpretation that links his (her) activity with all the scientists of the past. It is not so much the scientific method but as discussed by Hamming, it is a shared body of preconceptions that filter and color what we decide to label as data or facts. These preconceptions are formalized into a mathematical infrastructure that in addition to clarifying what is known, often disguises what is *not* known about classes of phenomena. These mathematical underpinnings are rarely called into question by the scientific community as a whole. It usually requires the actions of a heretic or iconoclast to show a scientific generation that their interpretation of the facts (the mathematical infrastructure) does not conform to physical (biological, social, etc.) reality. Perhaps the remarks of a thoughtful scientist with a philosophical turn, such as Wigner, who calls into question the efficacy of mathematics even when it does work, although more subtle, are of more lasting value. In spite of that, herein we focus on the consequent modifications in the formal infrastructure that are required to reconnect mathematical deductions and experimental observations when the two diverge. We do not linger on the failure of any particular model but rather, address those assumptions that are common to broad classes of mathematical models of physical phenomena.

The conventional wisdom of the physical sciences has been collected into an assortment of *traditional truths* which have become so obvious that they are, as we said, all but impossible to call into question. The first and perhaps foremost of these truths is that *physical theories are (and should be) quantitative*, as Lord Rutherford phrased it:

> All science is either physics or stamp collecting. Qualitative is nothing but poor quantitative.

The viewpoint was clarified by Thom[3] who points out that by the end of the seventeenth century, there were two main groups in physics, those that followed the physics of Descartes and those that followed the dictates of Newton[3]:

> Descartes, with his vortices, his hooked atoms, and the like, explained everything and calculated nothing. Newton, with the inverse square law of gravitation, calculated everything and explained nothing. History has endorsed Newton and relegated the Cartesian constructions to the domain of curious speculation.

This view is further elaborated by Bochner[4]:

> The demand of quantitativeness in physics seems to mean that every *specific* distinction, characterization, or determination of a state of a physical object and the transmission of specific knowledge and information, must *ultimately* be expressible in terms of real numbers, either single numbers or groupings of numbers, whether such numbers be given "intensively" through the medium of formulae or "extensively" through the medium of tabulation, graphs or charts.

Thus this first traditional truth takes the form that if it is not quantitative it is not scientific. This visceral belief has molded the science of the 20th century, in particular those emerging disciplines relating to life and the society in which we live. Much can be said of both the successes and failures of applying this dictum outside the physical sciences but a detailed review of this body of work would lead us far from the main purpose of this essay. Instead let us concentrate on the recent successes of applying certain *non-traditional truths* in counter-point to the traditional truths. The first of these non-traditional truths is that theories can be qualitative as well as quantitative. The most venerable proponent of this view in recent times was D'Arcy Thompson,[5] whose work in part motivated the development of catastrophe theory by René Thom.[3] Their interest in biological morphogenesis stimulated a new way of thinking about change – not the smooth, continuous quantitative change familiar in many physical phenomena, but abrupt, discontinuous, qualitative change familiar from the experience of "getting a joke", "having an insight", the breaking of a wave on the sea surface, the bursting of a bubble and so on.

Catastrophe theory has its foundation in topology and is therefore qualitative rather than quantitative, which is to say it deals with the forms of things and not with their magnitudes. This theory recognizes that many if not most interesting phenomena in nature involve discontinuities and it was designed to categorize their types. This approach stands in sharp contrast to the vast majority of available techniques which were designed for the quantitative study of continuous behavior.[6,7]

This is but one example of a mathematical discipline whose application to sciences emphasizes the qualitative over the quantitative. Another is the bifurcation behavior of nonlinear dynamic equations. A bifurcation is a quantitative change in the solution to a differential equation obtained through the variation of a control parameter. For example in certain systems manifesting a harmonic solution of unit period, changing the control parameter generates a sequence of subharmonic bifurcations in which the period of the solution doubles, doubles again and so on.[8] Eventually the solution winds up being irregular in time and such solutions have been used to model such apparently random phenomena as turbulent fluid flow. The aperiodic behavior of this final state has suggested a new paradigm for the unpredictable behavior of complex systems. This behavior is discussed further in Sec. 3.

It has been over half a century since the physics community last endured a complete shift in paradigm. That shift was, of course, the creation of quantum mechanics to describe the physics-of-the-very-small. A mathematical formalism developed in the 19th century to describe linear wave motion (Sturm-Liouville theory) discussed in Sec. 2, was applied by

Schrödinger to quantum mechanical wave functions and the first shot in the classical/quantum revolution was fired. The classical notion of determinism gave way to a probability amplitude – not even a proper classical probability, but a *probability amplitude.* Discreteness (quantization) in the very small required that we give up some of our preconceptions about predictability. Only quantities that are averaged in a particular way have any classical meaning, and the famous "correspondence principle" was invented by Bohr to save our intuition. The mathematics of classical physics could be used in quantum mechanics but not without a fundamental reinterpretation of the symbols. This process of reinterpretation has continued in an uninterrupted way until today and I have every confidence that it will continue through tomorrow. The revolution in the physics-of-the-very-small has been over for a long time, essentially being ended by the Copenhagen Conference at which time the present day interpretation of quantum phenomena was decided upon. Not all the contradictions and ambiguities of quantum theory were resolved at that conference, however. These difficulties are being addressed as part of the ongoing process of the evolution of quantum theory into a mature discipline. Even with all their initial mystery and counterintuitive results, quantum phenomena retain the fundamental truths. Quantum theory required the replacement of functions by operators and the introduction of probability amplitudes in order for the quantitative aspects of physics to survive, but that was a small price to pay.

The second fundamental truth that spans the macroscopic and microscopic is that *physical observables and their relationships are represented by analytic functions.* Since the time of Lagrange,[9] it has been accepted that mechanics and physics are described by smooth, continuous, unique functions, or at least piecewise continuous functions. A piecewise continuous function is one that has a finite number of discontinuities, e.g. the space-time trajectory of a billiard ball bouncing between the cushions of a pool table. (The turning points at the cushion are discontinuous.) This belief permeates physics because the evolution of physical processes are modeled by systems of dynamic equations and the solutions to such equations are thought to be continuous and differentiable on all but a finite number of points, that is to say the solutions are analytic functions.

There have always been a number of physical phenomena that could not be described by analytic functions, such as phase transitions and turbulence, to name two. In these phenomena, scaling is found to be crucial in the description of the underlying physics. Many scaling relations for complex systems in the physical, biological and psychosocial sciences involve noninteger exponents. The existence of noninteger exponents implies that such functions do not have a Taylor expansion and are therefore nonanalytic. One can interpret the noninteger exponents in these scaling relations as indications of singularities. The simple "laws" in physics, such as Ohm's law and the ideal gas law have integer exponents and are analytic. The simple alometric growth "laws" in botany and biology have noninteger exponents that are often irrational.[10] The traditional discussions of these alometric growth laws rely on simple physical arguments resulting in rational noninteger exponents, the observed deviation of the data from these values being assumed to be the result of experimental limitations.[11] More modern scaling arguments contend that these laws are often the result of renormalization group relations, which result from the fact that these systems have no fundamental scale.[12,13,14] These new ideas again require that we reinterpret existing data sets.

If the interpretation of a body of data is allowed to change, then where is the stability in science? Where is the deep and lasting understanding of the universe for which the scientist searches? Notice that both these questions contain the implicit assumption that truth is static; once discovered, it is unchanging over time. This kind of absolute truth is reminiscent of the notions of absolute space conceived by Galileo and absolute time embraced by Newton and a subsequent two centuries of scientists. The physics of Newton superceded the absolute space of Galileo and left only absolute time. Another paradigmatic shift, this time one due to Einstein, reintroduced the perspective of the observer into the interpretation of data, thereby nullifying the concept of the absolute in physics. This was, of course, relativity, but a similar dependence on the observer had also been made in quantum mechanics. Now we "know" that absolute space and absolute time are a fiction, and data coordinates must always be referenced back to the observer. But could such a situation also rise for truth? This question has an epistemological flavor and its answer emerges from a revolution that is now in progress in physics, to which we alluded earlier and which has direct and profound effects on each of the other sciences.

The third traditional truth is that *since the time evolution of physical systems are determined by systems of dynamic equations, if one completely specifies the initial state of the system then the solution to the dynamic equations determines the final state of that physical system.* That is to say that the final state can be *predicted* from a given initial state using the dynamic equations. Until recently, this view prompted scientists to casually assert that the evolution of these observables are absolutely predictable. While it is true that the Newtonian model is valid, it does *not* imply absolute predictability. This is not a quantum effect, but is rather a result of nonlinear dynamics system theory. This theory, certain aspects of which will be discussed subsequently, enables one to interpret the generic properties of complex systems without solving the equations of motion in detail. The particular results with which we will be concerned are the aperiodic (irregular) solutions mentioned earlier. These solutions are manifestations of the deterministic randomness that can arise from nonlinear interactions in the macroscopic world and is called *chaos.*[15] Since chaotic processes are irregular, they have limited predictability and call into question each of the first three traditional truths, i.e., whether the important aspects of a phenomena are quantitative, analytic and predictable.

We are accustomed to unpredictability in statistical physics in the context of complex systems having many degrees of freedom. For example, the vast number of molecules in a gas insures the validity of thermodynamic relations arising from averages over the molecular degrees of freedom. The classical view is that each of the microscopic degrees of freedom satisfy Newton's equations and that statistical fluctuations in macroscopic observables such as the pressure or mass density arise from uncertainty in the initial conditions of the unmeasured microscopic variables. Averaging over an ensemble of realizations of these unobserved initial states determines the macroscopic equations of motion which in turn determines the random behavior of the macroscopic observables. This is distinct from the chaos that can arise in nonlinear systems, having only a few macroscopic degrees of freedom.

Much time and effort has been devoted in classical statistical mechanics and more recently, in nonequilibrium statistical mechanics, to the derivation of macroscopic equations of motion from averages over the microscopic degrees of freedom. These activities have been motivated by the belief that the observed statistics are consistent with Newtonian determinism and are a consequence of a system's complexity rather than resulting from any intrinsic lack of predictability of the phenomena. Such systems are more properly described by distribution functions rather than vast numbers of microscopic equations of motion. Typical distributions that arise are the Poisson in which the average value determines all the moments of the distribution, and the Gaussian where the first and second moments are sufficient to determine all other moments.

These considerations lead to the fourth traditional truth; *physical systems can be characterized by fundamental scales of length, time and so on.* Such scales provide the meaningfulness of the fundamental units in the physical sciences without which measurements could not be made and quantification would not be possible. In the present context, these scales manifest themselves in the central moments of distributions as the average value of a physical observable and its variance. But what of distributions whose central moments diverge such as inverse power laws that are observed far and wide outside the physical sciences?[8,10] For example Zipf's law in linguistics concerning the relative frequency of word usage; Pareto's law in economics concerning the number of people at a given income level; Lotka's law in sociology concerning the number of publications by scientists, and so on. These hyperbolic distributions are known to have a statistical self-similarity resulting in the distribution density obeying a functional scaling law (renormalization group property). The renormalization group relation is the mathematical expression of the underlying process not having a fundamental scale. This lack of scale is similarly manifest in the divergence of the central moments. Classical scaling principles are based on the notion that the underlying process is uniform, filling an interval in a smooth, continuous fashion. The new principle is one that can generate richly detailed, heterogeneous, but self-similar structure at all scales. The non-traditional truth is then that *the structure of many systems are determined by the scale of the measuring instrument*, and such things as length are a function of the unit of measure, e.g., a fractal curve.[8,13,14]

Because of the complexity of real systems, each time one measures a given quantity, a certain amount of estimation is required, e.g., in a simple physics experiment, one estimates the markings on a ruler or the pointer on a gauge. Thus, if one measures a quantity X a given number of times, n say, then instead of having a single quantity x, one obtains a collection of quantities $\{x_1, x_2, \ldots, x_n\}$, often called an ensemble. T. Simpson (1755) (known for his rule in calculus) was the first scientist to recognize that the observed discrepancies between measurements follow a pattern that is characteristic of the ensemble. His observations were the forerunner of the *law of error* which asserts that there exists a relationship between the magnitude of an error and its frequency of occurrence in an ensemble. If we consider the relative number of times an error of a given magnitude occurs in a population of a given (large) size, i.e., the frequency of occurrence, we obtain an *estimate* of the probability an error of this order will occur.

In the continuous limit, i.e., the limit in which the number of independent observations of a quantity approaches infinity, the characteristics of any well-behaved measured quantity is specified by means of a distribution function. The general idea is that any particular measurement has little or no meaning in itself, it is only the collection of measurements, i.e., the ensemble, that has a physical interpretation and this meaning is manifest through the distribution function. The distribution function, also called the probability density, associates a probability with the occurrence of an event in the neighborhood of a given magnitude. This regularity in the form of the frequency distribution in measurements prompted the 19th century English eccentric Galton to write.[16]

> I know of scarcely anything so apt to impress the imagination as the wonderful form of cosmic order expressed by the "Law of Frequency of Error". The law would have been personified by the Greeks and deified, if they had known of it. It reigns with serenity and in complete self-effacement, amidst the wildest confusion. The huger the mob, and the greater the apparent anarchy, the more perfect is its sway. It is the supreme law of Unreason. Whenever a large sample of chaotic elements are taken in hand and marshalled in the order of their magnitude, an unsuspected and most beautiful form of regularity proves to have been latent all along.

He was, of course, referring to Gauss' law of error (1809) and the normal (Gaussian) distribution. In a statistical mechanics context, Gibbs used these arguments to aid in the development of the canonical distribution.

When distributions have sufficiently long tails, the first few moments will *not* characterize the distribution because they diverge. Distributions with *infinite* moments characterize processes described by non-integer exponents and surprises that run counter to our intuition. As mentioned, integer exponents can usually be traced back to the analytic behavior of an appropriate function which can be expanded in a Taylor series. Non-integer exponents imply the presence of singularities and the breakdown of a Taylor series due to the divergence of a coefficient. The main point is that singularities and thus non-integer exponents arise in complex systems because they exhibit structure on all scales. The geometric notion of a fractal carries over into the domain of statistics in that it is the fluctuations that are revealed at smaller and smaller scales as one magnifies the interval of observation.[14] Lévy studied the general properties of such processes in the 1920's and 1930's and generalized the Central Limit Theorem to include those for which the second moment diverged. Among the varied and interesting properties of the Lévy distribution (processes) is the fact that it obeys a scaling law, indicating that the statistical fluctuations in the underlying process maintain a self-similarity property.[17]

A principle that perhaps falls just short of a fundamental truth is that of *superposition*, but even so it is sufficiently pervasive that we treat it as the fifth traditional truth. It is worthwhile to note that the principle of superposition influenced development in the physical sciences and later dominated the thinking of scientists studying natural phenomena but also the behavior of society and the nature of life itself. The principle can be stated as follows: *a complex process can be decomposed into constituent elements, each element can be studied individually and reassembled to understand the whole.* In this view of the world, all phenomena can be understood by testing the effects of nonlinearity as perturbations, i.e., they are weak effects. This was certainly true of the law of errors just discussed, and just as surely *not* true of the irregular solutions of Newton's equations previously mentioned.

We now summarize the traditional truths that seem to permeate the science of the 20th century:

1. physical theories are and should be quantitative;
2. physical phenomena can by and large be represented by analytic functions;
3. the evolution of physical phenomena can be predicted from the equations of motion;
4. physical systems have fundamental scales; and
5. most phenomena satisfy the principle of superposition.

What is observed in the latter part of this century is a contraposing set of non-traditional truths:

1. qualitative theories are as important, and sometimes more important than quantitative ones;
2. many phenomena are singular in character and cannot be represented by analytic functions;
3. the evolution of many systems, although derivable from dynamic equations, are not necessarily predictable;

4. phenomena do not necessarily have a fundamental scale and are described by scaling relations; and
5. most phenomena violate the principle of superposition.

It is clear that just as there are interrelations among the traditional truths, so too there are interrelations among the non-traditional ones. The common themes in the latter case seem to be the lack of a fundamental scale and unpredictability.[8,14]

In Sec. 2, we briefly indicate how linearity has molded our scientific view of the world and restricted our interpretation of complex systems. In many cases, linear ideas have actually inhibited the understanding of certain phenomena since the proper (nonlinear) interpretation violates one's intuition. In Sec. 3, we discuss how the role of nonlinearity has been modifying our preconceptions about experimental results, that is to say, how our idea of traditional truths has been changing. This reorientation is a consequence of using a blend of topology and differential equations, i.e., nonlinear dynamics theory, in modeling natural systems. We emphasize that the nonlinear aspect of the system can totally destroy our ability to predict the system's evolution and in so doing forces us to re-examine the distinction between a deterministic system and a random one. We also explore some of the implications of the blurring together of these two concepts in the mathematical infrastructure of the physical sciences and elsewhere.

2. The Linear World View

Mathematical models of biological and social phenomena have traditionally relied on the paradigm of classical physics in the development of their mathematical formalisms. An example of this is the 1925 book of Lotka,[18] wherein much of the ground work was laid for the present day view of biophysics, mathematical biology, and so on. This work is a sterling example of the application of tightly reasoned physical and mathematical arguments to biological phenomena in the now traditional style. The potency of this paradigm lies in the ability of classical physics to relate cause and effect in physical phenomena through a sequence of formal implications and thereby, to make predictions. Not all natural phenomena are predictable however. Weather is an example of a physical phenomenon which remains unpredictable. Scientists believe that they understand how to construct the basic equations of motion governing the weather and to a greater or lesser extent, they understand how to solve these equations. But even with that, the weather remains an engima; predictions can only be made in terms of probabilities. The vulnerability of the classical physics paradigm is revealed in that these phenomena do not display a clear cause/effect relation. A slight perturbation in the equations of motion can generate an unpredictably large effect.[19] Thus we say that the underlying process is random and that the equations of motion are stochastic in the tradition of classical statistical mechanics. A great deal of scientific effort has gone into making this view consistent with the idea that the random elements in the description would disappear if sufficient information were available about the initial state of the system, so that *in principle* the evolution of the system is predictable. This point of view is a direct consequence of the linear view of physical phenomena and is inconsistent with the result of simple nonlinear models that have been used to describe the process.

The concept of linearity has played a central role in the development of models in all the sciences. It is only fairly recently that the importance of nonlinearities has intruded itself into the world of the working scientist.[8,15] Examples of phenomena that are dominated by nonlinearity that immediately come to mind are the changes in the weather over a few days' time,[19] the height of the next wave breaking on the beach as you sit in the hot sun, shivering from a cold wind blowing down your back, and the infuriating intermittency in the time intervals between the drips from the bathroom faucet just after you crawl into bed at night.[20] In some cases such as the weather, the phenomenon always *appears* to be random but in other cases such as the dripping faucet, sometimes the dripping is periodic and other times each drip appears to be independent of the preceding one, thereby forming an irregular sequence. Since what relates these phenomena is their nonlinearity, we must clarify the nature of this property.

Nonlinearity is one of those strange concepts that is defined by what it is not. As one physicist put it, "It is like having a zoo of non-elephants." Thus we need to clearly identify the properties of linearity in order to specify which property a particular nonlinear process does not have. Consider, for example, a complicated system that consists of a number of factors. One property of linearity is that the response of the action of each separate factor is proportional to its value. For example, the degree of scatter of balls on a pool table is in direct proportion to the impulse of the cue ball striking the triangular array. This is the *property of proportionality*. A second property is that the total response to an action is equal to the sum of the values of the separate factors. This is the *property of independence*. As one example of linearity we choose systems theory, where in the standard theory one asserts that a process (or system) is linear if the output of an operation is directly proportional to the input. The proportionality constant is a measure of the sensitivity of the system response to the input. Formally the response, R, of a physical system is linear when it is directly proportional to the applied force F. This relation can be expressed algebraically by the relation $R = \alpha F + \beta$, where α and β are constants. If there is no response in the absence of the applied force, then $\beta = 0$. In a linear system, if two distinct forces F_1 and F_2 are applied, the net response would be $R = \alpha_1 F_1 + \alpha_2 F_2$ where α_1 and α_2 are independent constants. If there are N independent applied forces denoted by the vector $\mathbf{F} = (F_1, F_2, \ldots, F_N)$ then the response of the system is linear if there is a vector $\boldsymbol{\alpha} = (\alpha_1, \alpha_2, \ldots, \alpha_N)$ of mutually independent constant components such that $R = \boldsymbol{\alpha} \cdot \mathbf{F} = \sum_{j=1}^{N} \alpha_j F_j$. In this last equation, we see that the total response of the system, here a scalar, is a sum of the independent applied forces F_j each weighted by its own sensitivity coefficient α_j. These ideas carry over to more general systems where $\mathbf{F}$ is a generalized time dependent force vector and $\mathbf{R}$ is the generalized response. For example, pushing a gyroscope yields a response that is directly proportional to the applied force but which acts in a direction at right angles to the direction of the applied force.

To understand the overwhelming importance of the linearity concept in the physical sciences, and later in the biological and social sciences, we briefly review how the traditional style of thought used in physics is tied to linearity. We begin with the first of the classical physicists, Sir Isaac Newton. To gain some insight into his view of the physical world, i.e., how he applied his equations of motion, we consider how he solved one of the outstanding physics questions of his day. The problem was to determine the speed of sound in air. Our focus is on how Newton, the physicist, thought about difficult problems for which he had no exact solution or indeed, no equation of motion and how the success that emerged from his style of thought (strategy of model construction) motivated future generations of scientists to think in the same way.

Recall that sound is a wave phenomenon and its description is given in elementary physics texts by a wave equation [cf. (2.5)]. However, the wave equation is a partial differential equation and such mathematical expressions had not been invented at the time Newton considered this problem (1686).[21] He did know that sound was a wave phenomenon because of the observations of such effects as reflection, refraction, and diffraction. Although the proper mathematics did not exist, Newton argued that a standing column of air could be modeled as a linear harmonic chain of equal-mass oscillators. The picture consisted of air molecules of equal masses arranged along a line and locally interacting with other air molecules by means of linear elastic forces to keep the molecules localized. A sound wave, after all, consists of periodic rarefactions and compressions in the air and is an ideal physical system for such a model.

The linear character of the model enabled Newton to reason to the solution of the equation of motion without explicitly writing it out. He was able to deduce that the speed of sound in air is $v = (p/\rho)^{1/2}$ where p is the ambient air pressure and ρ is the mass density of the column of air. Using the isothermal volume elasticity of air, i.e., the pressure itself, in his equation he obtained a speed of sound in air of 945 ft/s, a value 17 percent smaller than the observed value of 1142 ft/s.

Newton's argument was severely criticized by Lagrange (1759)[9] who presented a much more rigorous statement of the problem. Lagrange developed a simple mechanical model of an equivalent system, that of a string consisting of a chain of harmonic oscillators, which he represented by a set of ordinary differential equations, one for each spring. If ξ_n represents the displacement of the nth molecule from its equilibrium position, the equations he constructed are

$$\frac{d^2\xi_n}{dt^2} = \alpha(\xi_{n+1} - \xi_n) - \alpha(\xi_n - \xi_{n-1})\,;\, n = 1,\dots,N\,. \tag{2.1}$$

The n-particle is then linearly coupled to the $(n \pm 1)$-particles, i.e., its nearest neighbors, and the force causing this molecule to move is proportional to the net displacement of these three molecules from their rest positions. A crucial assumption in (2.1) is that the force between particles is linear in the oscillator displacements and only acts locally along the chain, i.e., the nth molecule moves in direct proportion to the activity of its nearest neighbors. This happens all down the chain described by the differential-difference equation (2.1), i.e., an equation that is continuous in time t but discretely indexed by its position along the chain n.

It is interesting to note that even with all his criticism and rigor, Lagrange obtained Newton's value for the speed of sound in air. This is how the matter stood until 1816 when Laplace (1825)[22] replaced the isothermal with the adiabatic elastic constant, arguing that the compressions and rarefactions of a longitudinal sound wave take place adiabatically. He obtained essentially exact agreement (within certain small error bounds) with experiment. Thus, the modeling of the complicated *continuum* process of sound propagation in air by a discrete linear harmonic chain was successful! *The success of the application of a harmonic chain to the description of a complicated physical process established a precedent that has evolved into the backbone of modeling in physics.*

The next steps after Newton in the formation of a linear world view were taken largely by a father and son team through the letter correspondence of John (father) and Daniel (son) Bernouli beginning in 1727. In their correspondence they established that a system of N point masses (in one dimension) has N independent modes of vibration, i.e., eigenfunctions and eigenfrequencies. The number of independent degrees of freedom of the system motion is equal to the number of entities being coupled together. The Bernoulli's reasoned that if you have "N" point masses then you can have N modes of vibration without making particular assumptions about how to construct the equations of motion as long as the coupling is weak. This was the first statement of the use of eigenvalues and eigenfunctions in physics. Simply stated, their conclusion means that one can take a complicated system consisting of "N" particles or N degrees of freedom and completely describe it by specifying one function per degree of freedom. If one knows what the eigenfunctions are (for waves, these are harmonic functions) and one knows what the eigenvalues are (again for waves, these are the frequencies) then one knows what the motion is. Thus, they had taken a complicated system and broken it up into readily identifiable and tractable pieces, each one being specified by an eigenfunction. If one knows the eigenfunctions and eigenvalues, then the contention was that one knows nearly everything about the system.

Later, Daniel (1755)[23] used this result to formulate the *principle of superposition*: the most general motion of a vibrating system is given by a linear superposition of its characteristic (eigen, proper) modes. The importance of this principle cannot be over-emphasized. As pointed out by Brillouin,[24] until that time all general statements in physics were concerned with mechanics, which is applicable to the dynamics of an individual particle. The principle of superposition was the first formulation of a *general law* pertaining to a system of particles. One might therefore date theoretical physics from the formulation of this principle. Note that the concepts of energy and momentum were still rather fragile things and their conservation laws had not yet been enunciated in 1755. Yet at that time, Daniel Bernouli was able to formulate this idea that the most general motion of a complicated system of particles is nothing more than a linear superposition of the motions of the constituent elements. That is a very powerful and pervasive point of view about how one can understand the evolution of a complex system.

This procedure was in fact used by Lagrange[9] in his study of the vibration of a string. Using the model (2.1) in which there are a finite number of harmonically bound identical-mass particles, he established the existence of a number of independent frequencies equal to the number of particles. This discrete system was then used to obtain the continuous wave equation by letting the number of particles become infinitely great while letting the masses of the individual particles vanish in such a way that the mass density of the string remained constant. The resulting frequencies of the continuum were found to be the harmonic frequencies of the stretched spring.

During the same period, the prolific theoretical physicist of that age, or any age, L. Euler, was investigating the vibrations of a continuous string (1748). In his analysis, he used a partial differential equation, as had Taylor some 30 years earlier, and was able to demonstrate for the first time that the transverse displacement of the string during a vibration can be expressed as an arbitrary function of $x \pm vt$, where x is the distance along the string, t is the time and v is the velocity of the vibration. Euler's solution had the form

$$u(x,t) = \sum_{n=1}^{\infty} \hat{f}(n) \cos n\pi t \sin n\pi x \,, \tag{2.2}$$

where the initial disturbance $f(x)$ has the form

$$f(x) = \sum_{n=1}^{\infty} \hat{f}(n) \sin n\pi x \,. \tag{2.3}$$

A similar result had been obtained by d'Alembert the preceding year (1747) in his discussion of the oscillation of a violin string. His (d'Alembert) proposed solution had the form

$$u(x,t) = \frac{1}{2} f(x+vt) + \frac{1}{2} f(x-vt) \,, \tag{2.4}$$

which can also be obtained from (2.2) by means of trigonometric identities for sines and cosines. Euler and d'Alembert introduced partial differential equations into the discussion of physical processes, in particular they studied solutions of the wave equation

$$\frac{\partial^2}{\partial t^2}\,\xi(x,t) = v^2\,\frac{\partial^2}{\partial x^2}\,\xi(x,t)\,. \tag{2.5}$$

This style of thought provided the theoretical underpinning for D. Bernoulli's insight into the principle of superposition.

Oddly enough Euler did not accept the principle of superposition. Since it was known that the characteristic modes of vibration were sine and cosine functions and since the general motion of the string was a function only of $x \pm vt$, it followed that the most general vibration of the string could be expressed as a series (superposition) of sines and cosines with arguments $x \pm vt$. Euler thought that superposition was fine as long as one was talking about individual waves that were real physical objects – say the waves on a transverse string so that one can take a superposition of them to give the overall displacement of the string. However, he thought, the idea of writing the *general motion* of a physical system as a superposition, the way D. Bernoulli contended, was unphysical, even though he was the first to show that you could write down the solutions to the wave equation that way.

You will no doubt recognize the above statement of the general motion (2.2) as a special case of Fourier's theorem; *any periodic function can be expressed as a series of sines and cosines of some fundamental period* T. His (Fourier) contribution to the linear world view began in 1807 in his study of the flow of heat in a solid body.[25] He established the principle of superposition for the heat equation. Note that the heat equation describes irreversible, dissipative processes unlike the wave equation which describes reversible periodic processes. It is significant that the principle of superposition could be applied to these two fundamentally different physical situations. Fourier (1822)[25] argued, and it was eventually proved rigorously by Dirichlet (1829),[26] that any piecewise continuous (smooth) function could be expanded as a trigonometric (Fourier) series, i.e., as sums of sines and cosines with proper weightings. Thus, the emerging disciplines of acoustics, heat transport, electromagnetic theory (Maxwell's equations) in the eighteenth and nineteenth centuries could all be treated in a unified way. The assumed linear nature of all these processes insured that they each satisfy the principle of superposition.

The penultimate form of the principle of superposition was expressed in the mathematical formulation of the Sturm-Louiville theory (1836–1838).[24] The authors of this theory demonstrated that a large class of differential equations (and thereby the associated physical problems intended to be modeled by these equations) could be cast in the form of an eigenvalue problem. The solution to the original equation could then be represented as a linear superposition of the characteristic or eigen-motions of the physics problem. This was immediately perceived as a systematic way of unravelling complex events in the physical world and giving them a relatively simple mathematical description. In the following century, this would allow the problems of the microscopic world to be treated in a mathematically rigorous way. All of quantum mechanics would become accessible due to the existence of this mathematical apparatus.

Thus at the turn of the century it was a commonly held belief in the physical sciences community that: *a complex process can be decomposed into constituent elements, each element can be studied individually and the interactive part deduced from reassembling the linear pieces.* This view formed the basis of macroscopic field theories and rested on the twin pillars of the Fourier decomposition of analytic functions and Sturm-Louiville theory. Physical reality could therefore be segmented; understood piecewise and superposed back again to form a new representation of the original system. This approach was to provide the context in which other physical problems were addressed. Moreover, all deviations from linearity in the interacting motions were assumed to be treatable by perturbation theory. Linear theory describes the dominant interaction and if there are any remaining pieces, e.g., the interactions are not just nearest neighbors or harmonic forces in a linear chain but involve anharmonic nonlinear interactions, they were assumed to be weak. Thus the philosophy was to solve the linear problem first, then treat the remaining interaction (that which was not treated quite properly) as a perturbation on the linear solution, assuming throughout that the perturbation is not going to modify things qualitatively. This is the whole notion of interpolating from a linear world view to the effects of nonlinearity. In the linear world view, all the effects of nonlinearity are perturbative (weak) effects.

Perhaps we can now better appreciate what it is we deny when we contend that this or that phenomenon is dominated by its nonlinear character. We abandon the *proportionate response* of the linear world view. A small change in the input does *not* imply a correspondingly small change in the output; instead a nonlinear system (process) may have a *disproportionate response* in the output, ranging from no effect (if the change is below some threshold value) to an overwhelming instability. The simple flick of a switch transforms the darkness of a room into light. What could be more mysterious or nonlinear?

3. Chaos and the World View

One of the most fruitful and brilliant ideas of the second half of the 1600's was the idea that the concept of function and the geometric representation of a curve are related. Geometrically, the notion of a linear relation between two quantities implies that if a graph is constructed with the ordinate denoting the values of one variable and the abscissa denoting the values of the other variable, then the relation in question appears as a straight line. The graph or curve lies in the "space" defined by the independent coordinate axes. In a dynamic system, the coordinate axis are defined by the possible values of the independent dynamic variables and its derivatives.

The state of a given system is defined by a point in the above space, often called either the state space or phase space of the system. As time moves on, the point traces out a curve, called an orbit or trajectory, which describes the history of the system's evolution. This geometrical representation of dynamics is one of the most useful tools in dynamic systems theory for analyzing the time-dependent properties of nonlinear systems. By nonlinear, we now know that we mean the output of the system is not proportional to the input. One implication of this is the following: If the system is linear, then two trajectories initiated at nearby points in phase space would evolve in close proximity, so that at any point in future time the two trajectories (and therefore the states of the system they represent) would also be near to one another. If the system is nonlinear then two such trajectories could diverge from one another and at subsequent times the trajectories could be arbitrarily far apart, i.e., the distance between the orbits does not evolve in a proportionate way. Of course this need not *necessarily* happen in a nonlinear system, it is a question of stability.

This brings us back to our recurrent example of the weather and to the question of its predictability. Its elusive nature has only recently come into sharper focus and made clear two distinct views of the character of the evolution of deterministic dynamic systems. These views were articulated by their respective proponents Laplace and Poincaré, writing nearly 100 years apart.

Laplace believed in strict determinism and to his mind, this implied complete predictability. Uncertainty for him is a consequence of imprecise knowledge, so that the probability theory is necessitated by incomplete and imperfect observations. This view is clearly expressed in the quotation[22]:

> The present state of the system of nature is evidently a consequence of what it was in the preceding moment and if we conceive of an intelligence which at a given instant comprehends all the relations of the entities of this universe, it could state the respective positions, motions, and general affects of all these entities at any time in the past or future.
>
> Physical astronomy, the branch of knowledge which does the greatest honor to the human mind, gives us an idea, albeit imperfect, of what such an intelligence would be. The simplicity of the law by which the celestial bodies move and the relations of their masses and distances, permit analysis to follow their motions up to a certain point; and in order to determine the state of the system of these great bodies in past or future centuries, it suffices for the mathematician that their position and their velocity be given by observation for any moment in time. Man owes that advantage to the power of the instrument he employs, and to the small number of relations that it embraces in its calculations. But ignorance of the different causes involved in the productions of events, as well as their complexity, taken together with the imperfection of analysis, prevents our reaching the same certainty about the vast majority of phenomena. Thus there are things that are uncertain for us, things more or less probable, and we seek to compensate for the impossibility of knowing them by determining their different degrees of likelihood. So it is that we owe to the weakness of the human mind one of the most delicate and ingenious of mathematical theories, the science of chance or probability.

Poincaré, on the other hand, sees an intrinsic inability to make predictions due to a sensitive dependence of the evolution of the system on the initial state of the system. He expressed this in the following way[27]:

> A very small cause which escapes our notice determines a considerable effect that we cannot fail to see, and then we say

> that the effect is due to chance. If we knew exactly the laws of nature and the situation of the universe at the initial moment, we could predict exactly the situation of the same universe at a suceeding moment. But even if it were the case that the natural laws had no longer any secret for us, we could still only know the initial situation with *the same approximation*, that is all we require, and we should say that the phenomenon had been predicted, that it is governed by laws. But it is not always so; it may happen that small differences in the initial conditions produce very great ones in the final phenomena. A small error in the former will produce an enormous error in the latter. Predictions becomes impossible, and we have the fortuitous phenomenon.

Thus we can clearly distinguish between the "traditional" view of Laplace and the "modern" view of Poincaré. The latter view is considered modern because we have now discovered that deterministic systems with only a few degrees of freedom can generate aperiodic behavior that for many purposes is indistinguishable from random fluctuations. We emphasize that the random aspect is fundamental to the system dynamics and gathering more information will not reduce the degree of uncertainty. Randomness generated in this way is now called *chaos.*[15] To understand chaos we need to discuss system dynamics a bit more.

We introduced the notion of a phase space and a trajectory to describe the dynamics of a system. Each choice of an initial state for the system produces a different trajectory. If, however, there is a limiting set in phase space to which all trajectories are drawn after a sufficiently long time, we say that the system dynamics are asymptotically described by an attractor. The attractor is the geometric limiting set on which all the trajectories eventually find themselves, i.e., the set of points in phase space to which the trajectories are attracted. Attractors come in many shapes and sizes, but they all have the property of occupying a finite volume of phase space. As a system evolves, it sweeps through the attractor, going through some regions rather rapidly and others quite slowly, but always staying on the attractor. Whether or not the system is chaotic is determined by how two initially adjacent trajectories cover the attractor over time. As Poincaré stated, if a small change in the initial separation of trajectories (error) produces an enormous change in their final separation, then the evolution is unpredictable. One question is how this growing separation, indicative of chaos, is accomplished on an attractor of finite size. The answer has to do with the layered structure necessary for an attractor to be chaotic.

Rossler[28] described chaos as resulting from the geometric operations of stretching and folding. Two initially nearby orbits cannot rapidly separate forever on a finite attractor, therefore the attractor must eventually fold over onto itself. Once folded, the attractor is again stretched and folded again. This process is repeated over and over, yielding an attractor structure with an infinite number of layers to be traversed by the various trajectories. The infinite richness of the attractor structure affords ample opportunity for trajectories to diverge and follow increasingly different paths. The finite size of the attractor insures that these diverging trajectories will eventually pass close to one another again, albeit on different layers of the attractor. Crutchfield *et al.*[15] visualize these orbits on a chaotic attractor as being shuffled by this process, much as a deck of cards is shuffled by a dealer. Thus the randomness of the chaotic orbits is a consequence of this shuffling process. This process of stretching and folding creates folds within folds ad infinitum resulting in the attractor often having a fractal structure in phase space.

The degree of irregularity (chaos) of the dynamic observable is closely related to the geometrical structure of the underlying attractor. In a chaotic attractor, this structure is often that of a fractal, which is neither smooth nor homogeneous and whose smaller-scale features are similar to its larger-scale form. There is no characteristic length scale associated with a fractal, so consequently there is none for the chaotic attractor. Mandelbrodt,[14] the father of fractals, argues that we need a new kind of geometry to study such structures. Euclidean geometry is concerned with the understanding of straight lines and regular forms and assumes that the world consists of continuous smooth curves in spaces of integer dimension. When we look at billowing cumulus clouds, trees of all kinds, coral formations and coastlines we observe that the notions of classical geometry are inadequate to describe them. Detail does *not* become less and less important as regions of these various structures are magnified, but perversely more and more detail is revealed at each level of magnification. The rich texture of these structures is characteristic of fractals, and a fractal geometry in non-integer dimensions is required to describe them.

There are a number of measures of the degree of chaos of these attractors. One is its *dimension*; integer values of the dimension indicate a simple attractor, non-integer dimension indicates a chaotic attractor in phase space. A second measure of the degree of irregularity generated by a chaotic attractor is the *entropy* of the motion. The entropy is interpreted as the average rate of stretching and folding of the attractor, or alternatively, as the average rate at which information is generated.[20] One can view the preparation of the initial state of the system as initializing a certain amount of information. The more precisely the initial state can be specified, the more information one has. This corresponds to localizing the initial state of the system in phase space. The amount of information is inversely proportional to the volume of state space localized by measurement. In a regular attractor, trajectories initiated in a given local volume stay near to one another as the system evolves, so the initial information is preserved in time and no new information is generated. Thus the initial information can be used to predict the final state of the system. In a chaotic attractor, the stretching and folding operations smear out the initial volume, thereby destroying the initial information as the system evolves and the dynamics create new information. Thus the initial uncertainty in the specification of the system is eventually smeared out over the entire attractor and all predictive power is lost, i.e., *all causal connection between the present and the future is lost.*[20]

Here we see the first crack in the foundation of the physics paradigm. This loss of predictability and by implication the loss of ability to establish causality calls into question the familiar process of model construction in science. Rosen[29] points out that the essential feature of a scientific model is the congruence between *causal processes* in the material world, and *implications* in their formal descriptions. Here we have assumed that a loss of predictability (implication) in the formal description carries an attendant loss of predictability in the natural world. This, however, need not be the case – it may be a limitation of the formal description that carries with it chaos rather than any intrinsic lack of predictability in the natural world. Either of these viewpoints necessitates a restructuring of interpretation of data sets, but in quite different ways. We return to this point subsequently.

A third measure of the degree of chaos associated with an attractor is the set of *Lyapunov exponents*. A number of scientists believe that the spectrum of Lyapunov exponents provides the most complete qualitative and quantitative characterization of chaotic behavior.[30] A system with one or more positive Lyapunov exponents is defined to be chaotic. The local stability properties of a system are determined by its response to perturbations; along certain directions the response can be stable (proportionate) whereas along others it can be unstable (disproportionate). If we consider a d-dimensional sphere of initial conditions and follow the evolution of this sphere in time, then in some directions the sphere will contract, whereas in others it will expand, thereby forming a d-dimensional ellipsoid. Thus, a d-dimensional system can be characterized by d exponents where the jth Lyapunov exponent quantifies the expansion or contraction of the flow along the jth ellipsoidal principal axis. The sum of the Lyapunov exponents is the average divergence, which for a dissipative system (possessing an attractor) must always be negative.

Consider a three dimensional phase space in which the limiting set (the attractor) can be characterized by the triple of Lyapunov exponents $(\lambda_1, \lambda_2, \lambda_3)$. The qualitative behavior of the attractor can be specified by determining the signs of the Lyapunov exponents only, i.e., (sign λ_1, sign λ_2, sign λ_3). As shown in Fig. 1a the triple $(-,-,-)$ corresponds to an attracting fixed point. In each of the three directions there is an exponential contraction of trajectories so that no matter what the initial state of the system, the trajectory will eventually wind up at the fixed point. This fixed point need not be the origin, as it would be for a dissipative linear system, but can be anywhere in phase space. The arrows shown in the figure do not necessarily represent trajectories since the fixed point can be approached at any angle by the evolving nonlinear system.

An attracting limit cycle is denoted by $(0,-,-)$ in which there are two contracting directions and one that is neutrally stable. In Fig. 1b, we see that this attractor resembles the orbit of a harmonic oscillator with a particular energy but that is not the case. The orbit of a harmonic oscillator or other conservative systems do not attract points from off the orbit onto itself. On the other hand, in a nonlinear dynamical system, an orbit has a basin of attraction so that all orbits initiated in the basin eventually wind up on the limit cycle.

The triple $(0,0,-)$ has two neutral directions and one that is contracting, so that the attractor is the 2-torus (doughnut) depicted in Fig. 1c. An example of such a system would be two coupled harmonic oscillators, where the two positions and two velocities of the oscillator would describe the dynamics. The constant energy (no dissipation) reduces the number of variables in this coupled system to three so that the system is described by the two constant radii and the two angles locating the trajectory on the surface of the torus.

Finally the combination $(+,0,-)$ corresponds to a chaotic attractor

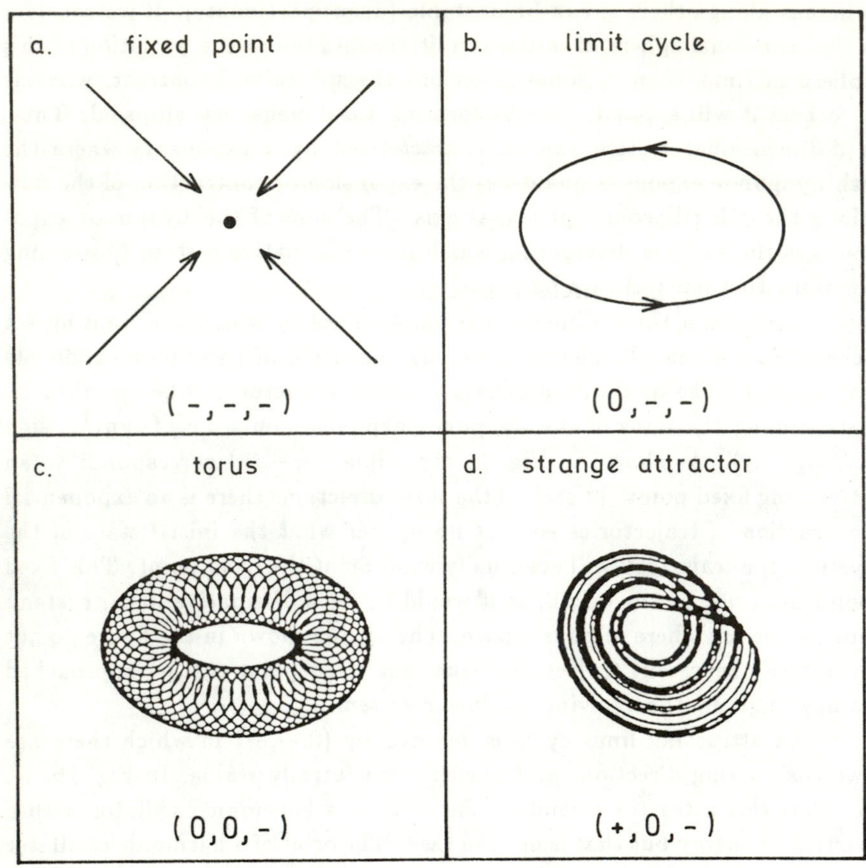

Fig. 1. The two-dimensional projection of simple attractors embedded in three dimensions are depicted. The signs of three Lyapunov characteristic exponents are shown for each attractor.

in which the trajectories expand in one direction, are neutrally stable in another and contracting in a third. In order for the trajectories to continuously expand in one direction and yet remain on a finite attractor, the attractor must undergo stretching and folding operations in this direction as discussed above. Figure 1d is the two-dimensional projection of the Rössler attractor.

The resolution of the apparent conflict between the traditional and the modern view of dynamic systems theory as presented in classical mechanics is that chaos is not inconsistent with the traditional notion of solving deterministic equations of evolution. As Ford states[31]

> ... Determinism means that Newtonian orbits exist and are unique, but since existence-uniqueness theorems are generally nonconstructive, they assert nothing about the character of the Newtonian orbits they define. Specifically, they do not preclude a Newtonian orbit from passing every computable test for randomness of being humanly indistinguishable from a realization of a truly random process. Thus, popular opinion to the contrary notwithstanding, there is absolutely no contradiction in the term "deterministically random". Indeed, it is quite reasonable to suggest that the most general definition of chaos should read: chaos means deterministically random ...

Thus if nonlinearities are ubiquitous then so too must be chaos. This led Ford to speculate on the existence of a generalized uncertainty principle based on the notion that the fundamental measures of physics are actually chaotic. The perfect clocks and meter sticks of Newton are replaced with "weakly interacting chaotic substitutes" so that the act of measurement itself introduces a small and uncontrollable error into the quantity being measured. Unlike the law of errors conceived by Gauss, which is based on linearity and the principle of superposition of independent events, the postulated errors arising from nonlinearities cannot be reduced by increasing the accuracy of one's measurements. The error (or noise) is generated by the intrinsic chaos associated with physical being.

In his unique style, Ford summarizes the situation in the following way[31]

> Although much, perhaps most, of man's impressive knowledge of the physical world is based on the analytic solution of dynamical systems which are integrable, such systems are, metaphorically speaking, as rare as integers on the real line. Of course, each integrable system is "surrounded"... by various other systems amenable to treatment by perturbation theory. But even in their totality, these systems form only an extremely small subset of the dynamical whole. If we depart from this small but precious oasis of analytically solvable, integrable or nearly integrable systems, we enter upon a vast desert wasteland of undifferentiated nonintegrability. There, in the trackless waste, we find the nomads: systems abandoned because they failed

> a qualifying test for integrability; systems exiled for exhibiting such complex behavior they were resistant to deterministic solution they were labeled intractable. Of course, we also find chaos in full residence everywhere...

So both the second traditional truth associated with analytic functions and the third traditional truth concerned with predictability need to be reconsidered and either modified or replaced.

There is another resolution to these conflicting points of view that in some ways is even more radical than admitting chaos into the ordered halls of physics. The second resolution is championed by Rosen[29] who points out the distinction between mathematical theories that are syntactic and those that are semantic. Both concepts were originally linguistic, the former dealing with what is true or meaningful in a language solely by virtue of form and pattern and the latter, the study of meaning or significance. The axiomatization of mathematics is an attempt to eliminate the semantic content so as to have no external referent in mathematical theory. Rosen argues that[29]:

> There are obvious parallels between the perceptions of mathematics as the manipulation of meaningless structureless, unanalyzable symbols according to fixed syntactic rules, and the perception of physical systems as collections of structureless, unanalyzable, elementary units ("atoms") manipulated by definite forces. Likewise, the formalistic belief that every mathematical truth can be generated in purely syntactic terms translates into *reductionism*, the belief that every physical truth translates into a truth about atoms and forces, and back again.

The modern formulation of the above idea of Natural Law is encapsulated in Fig. 2. Here we see that the events in the material world can be encoded into a formal system, implications can be made within the system and the results decoded back into the external world. That is to say that causal relations between events in the external world have a correspondence with implication relations in the formal world, in short $\boxed{1} = \boxed{2} + \boxed{3} + \boxed{4}$, indicating that the formalism *models* the material system.[29] As Rosen points out, Newtonian mechanics provided not only the implication structure in the formal system, but also indicated how the encoding and decoding is to be accomplished. Newton's modeling strategy was so successful that today the notion of a physical theory is almost always restricted to the study of the implication relations and ignores the process of encoding and decoding implicit in the modeling process.

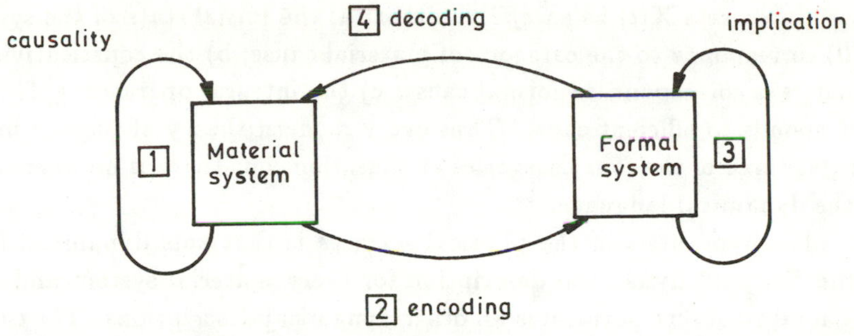

Fig. 2.

Rosen envisions that for any material system, there exists a "largest" dynamical description that captures all of the physical reality in a fully syntactical model, i.e., in a completely reductionist description. Following his arguments we note that the Newtonian revolution was thought to bring an end to Aristotelian science and to shift science from the study of the "why" of things to the "how" of things. The Aristotelian notion of *cause* was replaced by the weaker concept of *determinism*, which is embodied in the mathematical idea of a functional relation. The latter is a syntactical concept on the right-hand side of Fig. 2, whereas the former is not; it is a relation in the material system on the left-hand side of Fig. 2. Why then do scientists continue to use the notion of causality in their general discussions, and how are determinism and causality distinguished?

First let us recall that Aristotle had four categories of cause: 1) *material cause*, 2) *formal cause*, 3) *efficient cause*, 4) *final cause*. It is in fact only the final cause that modern science bans from physical models. In general a physical model consists of state $\mathbf{X}(t)$ and dynamic laws or equations of motion, i.e.,

$$\frac{d\mathbf{X}}{dt} = \mathbf{f}(\mathbf{X}, \boldsymbol{\alpha}), \tag{3.1}$$

where $\boldsymbol{\alpha}$ is a vector of (constitutive) parameters and $\mathbf{f}$ is derivable from the forces. Equation (3.1) can be formally integrated to obtain

$$\mathbf{X}(t) = \mathbf{X}(0) + \int_0^t \mathbf{f}(\mathbf{X}(\tau), \boldsymbol{\alpha}) d\tau . \tag{3.2}$$

Rosen interprets $\mathbf{X}(t)$ as an *effect* so that: a) the initial state of the system $\mathbf{X}(0)$ corresponds to the category of material cause; b) the constitutive parameters $\boldsymbol{\alpha}$ correspond to formal cause; c) the integral operator $\int_0^t \mathbf{f}(\cdot,\cdot)dt$ corresponds to efficient cause. Thus every modern theory of physics incorporates three of the four categories of causation, but there is no final cause in the dynamical language.[29]

The hypothesis in the physical sciences is that this dynamical form is the "largest" dynamical description for every material system and that its causal structure permeates all other dynamical descriptions. The causal structure in such systems is of an extremely simple type, so that Rosen calls systems that are exclusively describable in this way as *simple systems.* He refers to material systems which are not simple, in this sense, as *complex* and asserts that biology is repleat with complex systems. Thus complex systems are large enough to encompass final causation which is excluded from dynamical systems, implying that complex systems cannot be modeled as dynamic systems. A complex system is one in which the present state of the system can anticipate some future state and therefore be guided by it during the process of evolution. A complex system is therefore *anticipatory* rather than *reactive.* Of course one might argue that the notion of a basin of attraction, i.e., the set of initial conditions off the attractor which asymptotically wind up on the attractor, may have a teleological component in that the final set of states determines the evolution *from* the set of initial states. This, however, would be a generalization of what is meant by final cause in the traditional sense. Initial conditions in different regions of phase space would be attracted to different attractors just as different cells in biology grow into lungs, hearts, spleens, etc.

4. Further Ruminations

One of the questions the preceding discussion suggests is: Does chaos in a dynamical image of a material system possess an effective "inverse image" in a material system itself? An eloquent answer to this question was recently given to me by R. Rosen which I paraphrase below.

Rather than directly answering the original question, let us consider a second question. Namely, does there exist an infinite decimal, i.e., an irrational number such as π, in the material world? Irrational numbers are mathematically generic, that is to say typical, but we deal with their *practicality* by replacing them with rational numbers; with non-generic numbers. By doing so, we may be making errors (round-off errors), but if we are careful, and study our numerical analysis, we need never deal with the full reality of an infinite decimal. In fact, we can pretend that the real number system has no physical reality; the only thing that matters is the non-generic subset of rationals.

He went on to say that chaos is a way of magnifying the irrationality of numbers; making that irrationality visible, observable, and non-ignorable. So the two questions, the one regarding the 'reality' of chaos and the other the 'reality' of irrational numbers, are closely related. Chaos makes the replacement of an irrational (generic) number by a rational (non-generic) approximation untenable: it pushes the neglected infinite tail into the realm we have retained (given enough time). In fact, if chaos does exist in the material world, so must irrational numbers. Thus there must be irrational concentrations, irrational densities, and so on, regardless of our ability to measure them directly, or the resolving power of our meters. This does in fact appear to be the case as witnessed by the recent deluge of "observed" fractal phenomena.

The converse is not true, however. If we restrict ourselves to dynamics in which round-off errors do not accumulate, then we would never see chaos. Of course such dynamics are non-generic too. But this has been the habitat of theoretical physics since the time of Newton. Physics has so far always tacitly assumed that their systems are inherently non-generic. Closed systems, conservation laws, symmetry conditions, exact differential forms and a host of other assumptions are characteristic non-genericities imposed by physicists on the mathematical images of material reality which they allow, and in fact constitute the laws of nature which are claimed to be universal. Such restrictions are responsible for the predictive power which physics has enjoyed and physicists will not easily surrender them.

We close this essay by recalling Wigner's concern over the usefulness of mathematics in the physical sciences and our failure to understand why this is the case. The arguments that both he and Hamming used did not satisfy them, however it appears to me that one of Hammings' arguments is particularly compelling: *that we see what we look for.* In seeking to understand a given phenomenon, a scientist may make a sequence of measurements to refine the level of detail available about it; or perhaps observe how the event changes as the experimental parameters are changed. Another scientist may speculate on how the occurrence works by constructing a model of it and exploring the consequence of changing parameters in the model. When the two approaches give comparable results we observe that we have a model of the phenomenon. In the physical sciences such models are traditionally mathematical, as we have discussed. These models are then taught to would-be scientists who are encouraged to check them by conducting the appropriate experiments for themselves. Sure enough the experiments vindicate the suggested models. In this sense the novitiate scientist sees what he/she looks for. But more to the point the novice does not see the original encoding used to develop the model in the first place. The ability to directly apprehend the event has been compromised by learning the model. The mathematical infrastructure filters out those aspects of the phenomenon that are not understood.

Experimentalists might here remark that they are closer to "reality" because they are not shackled by any theoretical bias in conducting an experiment; the data stand outside the mathematical framework. This may be true with regard to any particular model, but even here the traditional truths are retained in the interpretation of the experimental results. The raw data are rarely published in a context-free form. But even when they are, the eye draws a curve through the sequence of data points (analytic function), the mind perceives clusterings in tables of numbers (important scales); the scientists work to find patterns and they do find them. Thus, although the experimentalists may not share the theorists' passion for mathematical modeling, they do share the quest for understanding events in accordance with natural law and physical principles. The traditional truths guide the hand of the experimentalist just as surely as they do the hand of the theoretician. It is only the emergence of contradictions that force either to re-examine the underpinnings of their activities.

The effects of nonlinearities presents such contradictions; the absence of a smallest scale in space and/or time in fractal processes and structures and the deterministic randomness of chaos – these things cry out for a new fundamental perspective in science. The apparent options open to us at a fundamental level are to abandon predictability and thereby causality due to the intrusion of chaos in our dynamic models or to relinquish the dynamic models themselves in terms of a more all encompassing mathematical representation of the material world. In either case the next decade or so should be a very interesting time to be a scientist, to find out how all of this turns out and how we learn to incorporate the disproportionate response into our understanding of the nature of things.

References

1. E. P. Wigner, *Commun. Pure Appl. Math.* **XIII** (1960) 1–14.
2. R. W. Hamming, *Am. Math. Mon.* **87** (1980) 81–90.
3. R. Thom, *Structural Stability and Morphogenesis* (Benjamin/Cummings Reading Mass, 1975).
4. S. Bochner, *The Role of Mathematics in the Rise of Science* (Princeton University Press, 1966).
5. D. W. Thompson, *On Growth and Form* (1915); abridged ed. (Cambridge, 1961).
6. A. Woodcock and M. Davis, *Catastrophe Theory* (E. P. Dutton, 1978).
7. P. T. Saunders, *An Introduction to Catastrophe Theory* (Cambridge University Press, 1980).
8. B. J. West, *An Essay on the Importance of Being Nonlinear* (Springer-Verlag, 1985).
9. J. L. Lagrange, *Recheserches sur la Nature et al Propagation du Son* (Miscellanea Taurinesiu, 1759).
10. N. MacDonald, *Trees and Networks in Biological Models* (John Wiley and Sons, 1983).
11. N. Rashevsky, *Mathematical Biophysics Physico-Mathematical Foundations of Biology* (Dover, 1960), vols. 1 and 2, 3rd rev. ed.
12. B. J. West, V. Bhargava and A. L. Goldberger, *J. Appl. Physiology* **60** (1986) 189–197.
13. B. J. West and A. L. Goldberger, *Am. Sci.* **75** (1987) 354–365.
14. B. B. Mandelbrodt, *The Fractal Geometry of Nature* (W. F. Freeman, 1982).
15. J. P. Crutchfield, J. D. Farmer, N. H. Packard and R. S. Shaw, *Chaos, Sci. Am.* **255** (1987) 46–57.
16. L. C. Tippett, *Sampling and Standard Error*, in *The World of Mathematics*, ed. J. R. Newman (Simon and Schuster, 1956), vol. 3, pp. 1459–1486.
17. E. W. Montroll and B. J. West, in *Fluctuation Phenomena*, eds. E. W. Montroll and J. L. Lebowitz (North Holland, 1987), 2nd ed.
18. A. J. Lotka, *Elements of Mathematical Biology* (Williams and Wilkins, 1925); reprint (Dover, 1956).
19. E. N. Lorenz, *J. Atmosph. Sci.* **20** (1963) 130.
20. R. Shaw, *The Dripping Faucet as a Model Chaotic System* (Areil Press, 1984).
21. I. Newton, *Principia*, Book II (1686).
22. P. S. Laplace, *Traité de Mécanique Celeste* (Paris, 1825).
23. D. Bernoulli, *Reflexions et Eclarrcissements sin les Nouvelles Vibrations des Cordes Exposes dans les Memoires de l'Academie, Roy. Acad.* (Berlin, 1755) 147.
24. L. Brillouin, *Wave Propagation in Periodic Structures* (Dover, 1946), 2nd ed.
25. J. Fourier, *Théorie Analytique do la Chaleu* (Paris, 1822).

26. Dirichlet, *J. Für Math.* **IV** (1829) 157.

27. H. Poincaré, *Chance* in *The World of Mathematics*, ed. J. R. Newman (Simon and Schuster, 1956), vol. 2.

28. O. E. Rössler, *Phys. Lett.* **57A** (1976) 397.

29. R. Rosen in *Proceedings of Brain Research, Artifical Intelligence, and Cognitive Science* (1986).

30. See, e.g., A. Wolf, J. B. Swift, H. L. Swinney and J. A. Vastano, *Physica* **16D** (1985) 285–317.

31. J. Ford, in *Directions in Chaos*, ed. Hou Bai-Lin (World Scientific, 1987).

The Unreasonable Effectiveness of Mathematics in the Natural Sciences*

Eugene P. Wigner
Department of Physics
Princeton University
Princeton, NJ 08540, USA

*Reprinted from *Commun. Pure Appl. Math.* XIII (1960) 1–14.

"and it is probable that there is some secret here which remains to be discovered." (C. S. Peirce)

There is a story about two friends, who were classmates in high school, talking about their jobs. One of them became a statistician and was working on population trends. He showed a reprint to his former classmate. The reprint started, as usual, with the Gaussian distribution and the statistician explained to his former classmate the meaning of the symbols for the actual population, for the average population, and so on. His classmate was a bit incredulous and was not quite sure whether the statistician was pulling his leg. "How can you know that?" was his query. "And what is this symbol here?" "Oh," said the statistician, "this is π." "What is that?" "The ratio of the circumference of the circle to its diameter." "Well, now you are pushing your joke too far," said the classmate, "surely the population has nothing to do with the circumference of the circle."

Naturally, we are inclined to smile about the simplicity of the classmate's approach. Nevertheless, when I heard this story, I had to admit to an eerie feeling because, surely, the reaction of the classmate betrayed only plain common sense. I was even more confused when, not many days later, someone came to me and expressed his bewilderment[1] with the fact that we make a rather narrow selection when choosing the data on which we test our theories. "How do we know that, if we made a theory which focusses its attention on phenomena we disregard and disregard some of the phenomena now commanding our attention, that we could not build another theory which has little in common with the present one but which, nevertheless, explains just as many phenomena as the present theory." It has to be admitted that we have no definite evidence that there is no such theory.

The preceding two stories illustrate the two main points which are the subjects of the present discourse. The first point is that mathematical concepts turn up in entirely unexpected connections. Moreover, they often permit an unexpectedly close and accurate description of the phenomena in these connections. Secondly, just because of this circumstance, and because we do not understand the reasons of their usefulness, we cannot know whether a theory formulated in terms of mathematical concepts is uniquely appropriate. We are in a position similar to that of a man who was provided

[1] The remark to be quoted was made by F. Werner when he was a student in Princeton.

with a bunch of keys and who, having open several doors in succession, always hit on the right key on the first or second trial. He became skeptical concerning the uniqueness of the coordination between keys and doors.

Most of what will be said on these questions will not be new; it has probably occurred to most scientists in one form or another. My principal aim is to illuminate it from several sides. The first point is that the enormous usefulness of mathematics in the natural sciences is something bordering on the mysterious and that there is no rational explanation for it. Second, it is just this uncanny usefulness of mathematical concepts that raises the question of the uniqueness of our physical theories. In order to establish the first point, that mathematics plays an unreasonably important role in physics, it will be useful to say a few words on the question "What is mathematics?", then, "What is physics?", then, how mathematics enters physical theories, and last, why the success of mathematics in its role in physics appears so baffling. Much less will be said on the second point: the uniqueness of the theories of physics. A proper answer to this question would require elaborate experimental and theoretical work which has not been undertaken to date.

What is Mathematics? Somebody once said that philosophy is the misuse of a terminology which was invented just for this purpose.[2] In the same vein, I would say that mathematics is the science of skillful operations with concepts and rules invented just for this purpose. The principal emphasis is on the invention of concepts. Mathematics would soon run out of interesting theorems if these had to be formulated in terms of the concepts which already appear in the axioms. Furthermore, whereas it is unquestionably true that the concepts of elementary mathematics and particularly elementary geometry were formulated to describe entities which are directly suggested by the actual world, the same does not seem to be true of the more advanced concepts, in particular the concepts which play such an important role in physics. Thus, the rules for operations with pairs of numbers are obviously designed to give the same results as the operations with fractions which we first learned without reference to "pairs of numbers". The rules for the operations with sequences, that is with irrational numbers, still belong to the category of rules which were determined so as to reproduce rules for the operations with quantities which were already

[2]This statement is quoted here from W. Dubislav's *Die Philosophie der Mathematik in der Gegenwart* (Junker und Dunnhaupt Verlag, 1932), p. 1.

known to us. Most more advanced mathematical concepts, such as complex numbers, algebras, linear operators, Borel sets – and this list could be continued almost indefinitely – were so devised that they are apt subjects on which the mathematician can demonstrate his ingenuity and sense of formal beauty. In fact, the definition of these concepts, with a realization that interesting and ingenious considerations could be applied to them, is the first demonstration of the ingenuity of the mathematician who defines them. The depth of thought which goes into the formation of the mathematical concepts is later justified by the skill with which these concepts are used. The great mathematician fully, almost ruthlessly, exploits the domain of permissible reasoning and skirts the impermissible. That his recklessness does not lead him into a morass of contradictions is a miracle in itself: certainly it is hard to believe that our reasoning power was brought, by Darwin's process of natural selection, to the perfection which it seems to possess. However, this is not our present subject. The principal point which will have to be recalled later is that the mathematician could formulate only a handful of interesting theorems without defining concepts beyond those contained in the axioms and that the concepts outside those contained in the axioms are defined with a view of permitting ingenious logical operations which appeal to our aesthetic sense both as operations and also in their results of great generality and simplicity.[3]

The complex numbers provide a particularly striking example for the foregoing. Certainly, nothing in our experience suggests the introduction of these quantities. Indeed, if a mathematician is asked to justify his interest in complex numbers, he will point, with some indignation, to the many beautiful theorems in the theory of equations, of power series and of analytic functions in general, which owe their origin to the introduction of complex numbers. The mathematician is not willing to give up his interest in these most beautiful accomplishments of his genius.[4]

What is Physics? The physicist is interested in discovering the laws of inanimate nature. In order to understand this statement, it is necessary

[3]M. Polanyi, in his ***Personal Knowledge*** (University of Chicago Press, 1958) says: "All these difficulties are but consequences of our refusal to see that mathematics cannot be defined without acknowledging its most obvious feature: namely, that it is interesting," (p. 188).

[4]The reader may be interested, in this connection, in Hilbert's rather testy remarks about intuitionism which "seeks to break up and to disfigure mathematics", *Abh. Math. Sem. Univ.* (Hamburg, 1922), Vol. 157, or Gesammelte Werke (Springer, 1935), p. 188.

to analyze the concept "law of nature".

The world around us is of baffling complexity and the most obvious fact about it is that we cannot predict the future. Although the joke attributes only to the optimist the view that the future is uncertain, the optimist is right in this case: the future is unpredictable. It is, as Schrödinger has remarked, a miracle that in spite of the baffling complexity of the world, certain regularities in the events could be discovered [1]. One such regularity, discovered by Galileo, is that two rocks, dropped at the same time from the same height, reach the ground at the same time. The laws of nature are concerned with such regularities. Galileo's regularity is a prototype of a large class of regularities. It is a surprising regularity for three reasons.

The first reason that it is surprising is that it is true not only in Pisa, and in Galileo's time, it is true everywhere on the Earth, was always true, and will always be true. This property of the regularity is a recognized invariance property and, as I had occasion to point out some time ago [2], without invariance principles similar to those implied in the preceding generalization of Galileo's observation, physics would not be possible. The second surprising feature is that the regularity which we are discussing is independent of so many conditions which could have an effect on it. It is valid no matter whether it rains or not, whether the experiment is carried out in a room or from the Leaning Tower, no matter whether the person who drops the rocks is a man or a woman. It is valid even if the two rocks are dropped, simultaneously and from the same height, by two different people. There are, obviously, innumerable other conditions which are all immaterial from the point of view of the validity of Galileo's regularity. The irrelevancy of so many circumstances which *could* play a role in the phenomenon observed, has also been called an invariance [2]. However, this invariance is of a different character from the preceding one since it cannot be formulated as a general principle. The exploration of the conditions which do, and which do not, influence a phenomenon is part of the early experimental exploration of a field. It is the skill and ingenuity of the experimenter which shows him phenomena which depend on a relatively narrow set of relatively easily realizable and reproducible conditions.[5] In the present case, Galileo's restriction of his observations to relatively heavy

[5] See, in this connection, the graphic essay of M. Deutsch (Daedalus, 1958), Vol. 87, p. 86. A. Shimony has called my attention to a similar passage in C. S. Peirce's *Essays in the Philosophy of Science* (The Liberal Arts Press, 1957), p. 237.

bodies was the most important step in this regard. Again it is true that if there were no phenomena which are independent of all but a manageably small set of conditions, physics would be impossible.

The preceding two points, though highly significant from the point of view of the philosopher, are not the ones which surprised Galileo most, nor do they contain a specific law of nature. The law of nature is contained in the statement that the length of time which it takes for a heavy object to fall from a given height is independent of the size, material and shape of the body which drops. In the framework of Newton's second "law", this amounts to the statement that the gravitational force which acts on the falling body is proportional to its mass but independent of the size, material and shape of the body which falls.

The preceding discussion is intended to remind, first, that it is not at all natural that "laws of nature" exist, much less that man is able to discover them.[6] The present writer had occasion, some time ago, to call attention to the succession of layers of "laws of nature", each layer containing more general and more encompassing laws than the previous one and its discovery constituting a deeper penetration into the structure of the universe than the layers recognized before [3]. However, the point which is most significant in the present context is that all these laws of nature contain, in even their remotest consequences, only a small part of our knowledge of the inanimate world. All the laws of nature are conditional statements which permit a prediction of some future events on the basis of the knowledge of the present, except that some aspects of the present state of the world, in practice the overwhelming majority of the determinants of the present state of the world, are irrelevant from the point of view of prediction. The irrelevancy is meant in the sense of the second point in the discussion of Galileo's theorem.[7]

As regards the present state of the world, such as the existence of the earth on which we live and on which Galileo's experiments were performed, the existence of the sun and of all our surroundings, the laws of nature are entirely silent. It is in consonance with this, first, that the laws of nature can be used to predict future events only under exceptional circumstances

[6] E. Schrödinger, in his *What is Life* (Cambridge University Press, 1945) says that this second miracle may well be beyond human understanding (p. 31).

[7] The writer feels sure that it is unnecessary to mention that Galileo's theorem, as given in the text, does not exhaust the content of Galileo's observations in connection with the laws of freely falling bodies.

– when all the relevant determinants of the present state of the world are known. It is also in consonance with this that the construction of machines, the functioning of which he can foresee, constitutes the most spectacular accomplishment of the physicist. In these machines, the physicist creates a situation in which all the relevant coordinates are known so that the behavior of the machine can be predicted. Radars and nuclear reactors are examples of such machines.

The principal purpose of the preceding discussion is to point out that the laws of nature are all conditional statements and they relate only to a very small part of our knowledge of the world. Thus, classical mechanics, which is the best known prototype of a physical theory, gives the second derivatives of the positional coordinates of all bodies, on the basis of the knowledge of the positions, etc., of these bodies. It gives no information on the existence, the present positions, or velocities of these bodies. It should be mentioned, for the sake of accuracy, that we have learned about thirty years ago that even the conditional statements cannot be entirely precise: that the conditional statements are probability laws which enable us only to place intelligent bets on future properties of the inanimate world, based on the knowledge of the present state. They do not allow us to make categorical statements, not even categorical statements conditional on the present state of the world. The probabilistic nature of the "laws of nature" manifests itself in the case of machines also, and can be verified, at least in the case of nuclear reactors, if one runs them at very low power. However, the additional limitation of the scope of the laws of nature[8] which follows from their probabilistic nature, will play no role in the rest of the discussion.

The Role of Mathematics in Physical Theories. Having refreshed our minds as to the essence of mathematics and physics, we should be in a better position to review the role of mathematics in physical theories.

Naturally, we do use mathematics in everyday physics to evaluate the results of the laws of nature, to apply the conditional statements to the particular conditions which happen to prevail or happen to interest us. In order that this be possible, the laws of nature must already be formulated in mathematical language. However, the role of evaluating the consequences of already established theories is not the most important role of mathematics in physics. Mathematics, or, rather, applied mathematics, is not so much the master of the situation in this function: it is merely serving as a tool.

[8] See, for instance, E. Schrödinger, reference [1].

Mathematics does play, however, also a more sovereign role in physics. This was already implied in the statement, made when discussing the role of applied mathematics, that the laws of nature must be already formulated in the language of mathematics to be an object for the use of applied mathematics. The statement that the laws of nature are written in the language of mathematics was properly made three hundred years ago[9]; it is now more true than ever before. In order to show the importance which mathematical concepts possess in the formulation of the laws of physics, let us recall, as an example, the axioms of quantum mechanics as formulated, explicitly, by the great mathematician, von Neumann, or, implicitly, by the great physicist, Dirac [4,5]. There are two basic concepts in quantum mechanics: states and observables. The states are vectors in Hilbert space, the observables self-adjoint operators on these vectors. The possible values of the observations are the characteristic values of the operators – but we had better stop here lest we engage in a listing of the mathematical concepts developed in the theory of linear operators.

It is true, of course, that physics chooses certain mathematical concepts for the formulation of the laws of nature, and surely only a fraction of all mathematical concepts is used in physics. It is true also that the concepts which were chosen were not selected arbitrarily from a listing of mathematical terms but were developed, in many if not most cases, independently by the physicist and recognized then as having been conceived before by the mathematician. It is not true, however, as is so often stated, that this had to happen because mathematics uses the simplest possible concepts and these were bound to occur in any formalism. As we saw before, the concepts of mathematics are not chosen for their conceptual simplicity – even sequences of pairs of numbers are far from being the simplest concepts – but for their amenability to clever manipulations and to striking, brilliant arguments. Let us not forget that the Hilbert space of quantum mechanics is the complex Hilbert space, with a Hermitean scalar product. Surely to the unpreoccupied mind, complex numbers are far from natural or simple and they cannot be suggested by physical observations. Furthermore, the use of complex numbers is in this case not a calculational trick of applied mathematics but comes close to being a necessity in the formulation of the laws of quantum mechanics. Finally, it now begins to appear that not only numbers but so-called analytic functions are destined

[9] It is attributed to Galileo.

to play a decisive role in the formulation of quantum theory. I am referring to the rapidly developing theory of dispersion relations.

It is difficult to avoid the impression that a miracle confronts us here, quite comparable in its striking nature to the miracle that the human mind can string a thousand arguments together without getting itself into contradictions or to the two miracles of the existence of laws of nature and of the human mind's capacity to divine them. The observation which comes closest to an explanation for the mathematical concepts' cropping up in physics which I know is Einstein's statement that the only physical theories which we are willing to accept are the beautiful ones. It stands to argue that the concepts of mathematics, which invite the exercise of so much wit, have the quality of beauty. However, Einstein's observation can at best explain properties of theories which we are willing to believe and has no reference to the intrinsic accuracy of the theory. We shall, therefore, turn to this latter question.

Is the Success of Physical Theories Truly Surprising? A possible explanation of the physicist's use of mathematics to formulate his laws of nature is that he is a somewhat irresponsible person. As a result, when he finds a connection between two quantities which resembles a connection well-known from mathematics, he will jump at the conclusion that the connection *is* that discussed in mathematics simply because he does not know of any other similar connection. It is not the intention of the present discussion to refute the charge that the physicist is a somewhat irresponsible person. Perhaps he is. However, it is important to point out that the mathematical formulation of the physicist's often crude experience leads in an uncanny number of cases to an amazingly accurate description of a large class of phenomena. This shows that the mathematical language has more to commend it than being the only language which we can speak; it shows that it is, in a very real sense, the correct language. Let us consider a few examples.

The first example is the oft-quoted one of planetary motion. The laws of falling bodies became rather well-established as a result of experiments carried out principally in Italy. These experiments could not be very accurate in the sense in which we understand accuracy today partly because of the effect of air resistance and partly because of the impossibility, at that time, to measure short time intervals. Nevertheless, it is not surprising that as a result of their studies, the Italian natural scientists acquired a familiarity with the ways in which objects travel through the atmosphere.

It was Newton who then brought the law of freely falling objects into relation with the motion of the moon, noted that the parabola of the thrown rock's path on the earth, and the circle of the moon's path in the sky, are particular cases of the same mathematical object of an ellipse and postulated the universal law of gravitation, on the basis of a single, and at that time very approximate, numerical coincidence. Philosophically, the law of gravitation as formulated by Newton was repugnant to his time and to himself. Empirically, it was based on very scanty observations. The mathematical language in which it was formulated contained the concept of a second derivative and those of us who have tried to draw an osculating circle to a curve know that the second derivative is not a very immediate concept. The law of gravity which Newton reluctantly established and which he could verify with an accuracy of about 4% has proved to be accurate to less than a ten thousandth of a per cent and became so closely associated with the idea of absolute accuracy that only recently did physicists become again bold enough to inquire into the limitations of its accuracy.[10] Certainly, the example of Newton's law, quoted over and over again, must be mentioned first as a monumental example of a law formulated in terms which appear simple to the mathematician, which has proved accurate beyond all reasonable expectation. Let us just recapitulate our thesis on this example: first, the law, particularly since a second derivative appears in it, is simple only to the mathematician, not to common sense or to non-mathematically-minded freshmen; second, it is a conditional law of very limited scope. It explains nothing about the earth which attracts Galileo's rocks, or about the circular form of the moon's orbit, or about the planets of the sun. The explanation of these initial conditions is left to the geologist and the astronomer, and they have a hard time with them.

The second example is that of ordinary, elementary quantum mechanics. This originated when Max Born noticed that some rules of computation, given by Heisenberg, were formally identical with the rules of computation with matrices, established a long time before by mathematicians. Born, Jordan and Heisenberg then proposed to replace by matrices the position and momentum variables of the equations of classical mechanics [6]. They applied the rules of matrix mechanics to a few highly idealized problems and the results were quite satisfactory. However, there was, at that time, no rational evidence that their matrix mechanics would prove

[10]See, for instance, R. H. Dicke *American Scientist* (1959), Vol. 25.

correct under more realistic conditions. Indeed, they say "if the mechanics as here proposed should already be correct in its essential traits." As a matter of fact, the first application of their mechanics to a realistic problem, that of the hydrogen atom, was given several months later, by Pauli. This application gave results in agreement with experience. This was satisfactory but still understandable because Heisenberg's rules of calculation were abstracted from problems which included the old theory of the hydrogen atom. The miracle occurred only when matrix mechanics, or a mathematically equivalent theory, was applied to problems for which Heisenberg's calculating rules were meaningless. Heisenberg's rules presupposed that the classical equations of motion had solutions with certain periodicity properties; and the equations of motion of the two electrons of the helium atom, or of the even greater number of electrons of heavier atoms, simply do not have these properties, so that Heisenberg's rules cannot be applied to these cases. Nevertheless, the calculation of the lowest energy level of helium, as carried out a few months ago by Kinoshita at Cornell and by Bazley at the Bureau of Standards, agree with the experimental data within the accuracy of the observations, which is one part in ten millions. Surely in this case we "got something out" of the equations that we did not put in.

The same is true of the qualitative characteristics of the "complex spectra", that is the spectra of heavier atoms. I wish to recall a conversation with Jordan who told me, when the qualitative features of the spectra were derived, that a disagreement of the rules derived from quantum mechanical theory, and the rules established by empirical research, would have provided the last opportunity to make a change in the framework of matrix mechanics. In other words, Jordan felt that we would have been, at least temporarily, helpless had an unexpected disagreement occurred in the theory of the helium atom. This was, at that time, developed by Kellner and by Hilleraas. The mathematical formalism was too clear and unchangeable so that, had the miracle of helium which was mentioned before not occurred, a true crisis would have arisen. Surely, physics would have overcome that crisis in one way or another. It is true, on the other hand, that physics as we know it today would not be possible without a constant recurrence of miracles similar to the one of the helium atom which is perhaps the most striking miracle that has occurred in the course of the development of elementary quantum mechanics, but by far not the only one. In fact, the number of analogous miracles is limited, in our view, only by our willingness to go after more similar ones. Quantum mechanics had, nevertheless many almost equally striking successes which gave us the firm conviction that it is, what we call, correct.

The last example is that of quantum electrodynamics, or the theory of the Lamb shift. Whereas Newton's theory of gravitation still had obvious connections with experience, experience entered the formulation of matrix mechanics only in the refined or sublimated form of Heisenberg's prescriptions. The quantum theory of the Lamb shift, as conceived by Bethe and established by Schwinger, is a purely mathematical theory and the only direct contribution of experiment was to show the existence of a measurable effect. The agreement with calculation is better than one part in a thousand.

The preceding three examples, which could be multiplied almost indefinitely, should illustrate the appropriateness and accuracy of the mathematical formulation of the laws of nature in terms of concepts chosen for their manipulability, the "laws of nature" being of almost fantastic accuracy but of strictly limited scope. I propose to refer to the observation which these examples illustrate as the empirical law of epistemology. Together with the laws of invariance of physical theories, it is an indispensable foundation of these theories. Without the laws of invariance the physical theories could have been given no foundation of fact; if the empirical law of epistemology were not correct, we would lack the encouragement and reassurance which are emotional necessities without which the "laws of nature" could not have been successfully explored. Dr. R. G. Sachs, with whom I discussed the empirical law of epistemology, called it an article of faith of the theoretical physicist, and it is surely that. However, what he called our article of faith can be well supported by actual examples – many examples in addition to the three which have been mentioned.

The Uniqueness of the Theories of Physics. The empirical nature of the preceding observation seems to me to be self-evident. It surely is not a "necessity of thought" and it should not be necessary, in order to prove this, to point to the fact that it applies only to a very small part of our knowledge of the inanimate world. It is absurd to believe that the existence of mathematically simple expressions for the second derivative of the position is self-evident, when no similar expressions for the position itself or for the velocity exist. It is therefore surprising how readily the wonderful gift contained in the empirical law of epistemology was taken for granted. The ability of the human mind to form a string of 1000 conclusions and still remain "right", which was mentioned before, is a similar gift.

Every empirical law has the disquieting quality that one does not know its limitations. We have seen that there are regularities in the events in the world around us which can be formulated in terms of mathematical concepts with an uncanny accuracy. There are, on the other hand, aspects of the world concerning which we do not believe in the existence of any accurate regularities. We call these initial conditions. The question which present itself is whether the different regularities, that is the various laws of nature which will be discovered, will fuse into a single consistent unit, or at least asymptotically approach such a fusion. Alternately, it is possible that there always will be some laws of nature which have nothing in common with each other. At present, this is true, for instance, of the laws of heredity and of physics. It is even possible that some of the laws of nature will be in conflict with each other in their implications, but each convincing enough in its own domain so that we may not be willing to abandon any of them. We may resign ourselves to such a state of affairs or our interest in clearing up the conflict between the various theories may fade out. We may lose interest in the "ultimate truth", that is in a picture which is a consistent fusion into a single unit of the little pictures, formed on the various aspects of nature.

It may be useful to illustrate the alternatives by an example. We now have, in physics, two theories of great power and interest: the theory of quantum phenomena and the theory of relativity. These two theories have their roots in mutually exclusive groups of phenomena. Relativity theory applies to macroscopic bodies, such as stars. The event of coincidence, that is in ultimate analysis of collision, is the primitive event in the theory of relativity and defines a point in space-time, or at least would define a point if the colliding particles were infinitely small. Quantum theory has it roots in the microscopic world and, from its point of view, the event of coincidence, or of collision, even if it takes place between particles of no spatial extent, is not primitive and not at all sharply isolated in space-time. The two theories operate with different mathematical concepts – the four dimensional Riemann space and the infinite dimensional Hilbert space, respectively. So far, the two theories could not be united, that is, no mathematical formulation exists to which both of these theories are approximations. All physicists believe that a union of the two theories is inherently possible and that we shall find it. Nevertheless, it is possible also to imagine that no union of the two theories can be found. This example illustrates the two possibilities, of union and of conflict, mentioned before, both of which are conceivable.

In order to obtain an indication as to which alternative to expect ultimately, we can pretend to be a little more ignorant than we are and place ourselves at a lower level of knowledge than we actually possess. If we can find a fusion of our theories on this lower level of intelligence, we can confidently expect that we will find a fusion of our theories also at our real level of intelligence. On the other hand, if we would arrive at mutually contradictory theories at a somewhat lower level of knowledge, the possibility of the permanence of conflicting theories cannot be excluded for ourselves either. The level of knowledge and ingenuity is a continuous variable and it is unlikely that a relatively small variation of this continuous variable changes the attainable picture of the world from inconsistent to consistent.[11]

Considered from this point of view, the fact that some of the theories which we know to be false give such amazingly accurate results, is an adverse factor. Had we somewhat less knowledge, the group of phenomena which these "false" theories explain, would appear to us to be large enough to "prove" these theories. However, these theories are considered to be "false" by us just for the reason that they are, in ultimate analysis, incompatible with more encompassing pictures and, if sufficiently many such false theories are discovered, they are bound to prove also to be in conflict with each other. Similarly, it is possible that the theories which we consider to be "proved" by a number of numerical agreements which appears to be large enough for us, are false because they are in conflict with a possible more encompassing theory which is beyond our means of discovery. If this were true, we would have to expect conflicts between our theories as soon as their number grows beyond a certain point and as soon as they cover a sufficiently large number of groups of phenomena. In contrast to the article of faith of the theoretical physicist mentioned before, this is the nightmare of the theorist.

Let us consider a few examples of "false" theories which give, in view

[11]This passage was written after a great deal of hesitation. The writer is convinced that it is useful, in epistemological discussions, to abandon the idealization that the level of human intelligence has a singular position on an absolute scale. In some cases it may even be useful to consider the attainment which is possible at the level of the intelligence of some other species. However, the writer also realizes that his thinking along the lines indicated in the text was too brief and not subject to sufficient critical appraisal to be reliable.

of their falseness, alarmingly accurate descriptions of groups of phenomena. With some goodwill, one can dismiss some of the evidence which these examples provide. The success of Bohr's early and pioneering ideas on the atom was always a rather narrow one and the same applies to Ptolemy's epicycles. Our present vantage point gives an accurate description of all phenomena which these more primitive theories can describe. The same is not true any more of the so-called free-electron theory which gives a marvellously accurate picture of many, if not most, properties of metals, semiconductors and insulators. In particular, it explains the fact, never properly understood on the basis of the "real theory", that insulators show a specific resistance to electricity which may be 10^{26} times greater than that of metals. In fact, there is no experimental evidence to show that the resistance is not infinite under the conditions under which the free-electron theory would lead us to expect an infinite resistance. Nevertheless, we are convinced that the free-electron theory is a crude approximation which should be replaced in the description of all phenomena concerning solids, by a more accurate picture.

If viewed from our real vantage point, the situation presented by the free-electron theory is irritating but is not likely to forebode any inconsistencies which are unsurmountable for us. The free-electron theory raises doubts as to how much we should trust numerical agreement between theory and experiment as evidence for the correctness of the theory. We are used to such doubts.

A much more difficult and confusing situation would arise if we could, some day, establish a theory of the phenomena of consciousness, or of biology, which would be as coherent and convincing as our present theories of the inanimate world. Mendel's laws of inheritance and the subsequent work on genes may well form the beginning of such a theory as far as biology is concerned. Furthermore, it is quite possible that an abstract argument can be found which shows that there is a conflict between such a theory and the accepted principles of physics. The argument could be of such abstract nature that it might not be possible to resolve the conflict, in favor of one or of the other theory, by an experiment. Such a situation would put a heavy strain on our faith in our theories and on our belief in the reality of the concepts which we form. It would give us a deep sense of frustration in our search for what I called the "ultimate truth". The reason that such a situation is conceivable is that, fundamentally, we do not know why our theories work so well. Hence their accuracy may not prove their truth and consistency. Indeed, it is this writer's belief that something rather akin to the situation which was described above exists if the present laws of heredity and of physics are confronted.

Let me end on a more cheerful note. The miracle of the appropriateness of the language of mathematics for the formulation of the laws of physics is a wonderful gift which we neither understand nor deserve. We should be grateful for it and hope that it will remain valid in future research and that it will extend, for better or for worse, to our pleasure even though perhaps also to our bafflement, to wide branches of learning.

The writer wishes to record here his indebtedness to Dr. M. Polanyi who, many years ago, deeply influenced his thinking on problems of epistemology, and to V. Bargmann whose friendly criticism was material in achieving whatever clarity was achieved. He is also greatly indebted to A. Shimony for reviewing the present article and calling his attention to C. S. Peirce's papers.

Bibliography

[1] Schrödinger, E., *Uber Indeterminismus in der Physik*, J. A. Barth, Leipzig, 1932; also Dubislav, W., *Naturphilosophie*, Junker and Dünnhaupt, Berlin, 1933, Chap. 4.

[2] Wigner, E. P., *Invariance in phusical theory*, Proc. Amer. Philos. Soc., Vol. **93**, 1949, pp. 521–526.

[3] Wigner, E. P., *The limits of science*, Proc. Amer. Philos. Soc., Vol. 94, 1950, p. 422 also Margenau, H., *The Nature of Physical Realisty*. McGraw-Hill, New York, 1950, Chap. 8.

[4] Dirac, P. A. M., *Quantum Mechanics*, 3rd Edit., Clarendon Press, Oxford, 1947.

[5] von Neumann, J., *Mathematische Grundlagen der Quantenmechanik*, Springer, Berlin, 1932. English translation, Princeton Univ. Press, 1955.

[6] Born, M., and Jordan, P., *On quantum mechanics*, Zeits. f. Physik, No. **34**, 1925, pp. 858–888. Born, M., Heisenberg, W., and Jordan, P., *On quantum mechanics, Part II*, Zeits. f. Physik, No. **35**, 1926, pp. 557–615. (The quoted sentences occurs in the latter article, p. 558.)

The Effectiveness of Mathematics in Fundamental Physics

A. Zee
Institute for Theoretical Physics
University of California
Santa Barbara
CA 93106, USA

When I was a kid, so to speak, I came across and read Eugene Wigner's article[1] on "The Unreasonable Effectiveness of Mathematics in the Natural Sciences". I was impressed. Thus, I was rather pleased when Ron Mickens asked me, this many years later, to write a similar article.

Upon re-reading Wigner's article recently, I realized that Wigner had precious little to say, and perhaps necessarily so. I can hardly imagine that anyone could have much substantive to say about the question implied by the title of Wigner's article. Some of my colleagues may feel that, as a working theoretical physicist, I should not even be thinking about this question. But surely part of the fun of being a theoretical physicist is to be able to muse about such questions.

Furthermore, thirty or so years have passed since Wigner's article was published, and what we think of as physics and mathematics have shifted, at least in focus. It may be worthwhile to look at the question again. I want to say right from the start that I have neither a coherent theme nor a startling insight. Rather, I can offer only a disconnected series of observations, anecdotes, and musings. Anyway, what follows was known in my college days as "shooting the breeze".

Wigner's contribution lies, of course, in raising the question in the first place. In connection with Wigner's ability to ask seemingly profound questions, I may perhaps tell a little story. At around the time when I first read Wigner's article, I was an undergraduate. One winter day, I had to go back to the physics department after dinner to do some work. While I was eating dinner, it started to snow heavily outside. Trudging over to the physics building, I slipped and fell a couple of times. By the time I got there, I was literally covered with snow. As I staggered into the building, Eugene Wigner, with his heavy overcoat and hat on, was just about to go out. He looked at me carefully, and then he asked me, in that strangely solemn way that he had, "Excoose me, pleeze, izit snowing outside?"

Now surely that was a truly profound question if I ever heard one. The unreasonable effectiveness of visual inspection in comprehending reality? But while I did battle with such deep thoughts, Wigner had already walked out and disappeared into the howling storm.

A physics colleague once remarked to me that questions such as whether the effectiveness of mathematics in the natural sciences is reasonable or not have the curious property of being either incredibly profound or incredibly trivial. I am inclined to think that they are profound. Leaving that aside, let us try to understand and define the words in Wigner's question.

Effectiveness? What do we compare mathematics with in the effectiveness sweepstake? The trouble is that any alternative we can think of is so utterly ineffective by comparison. But how do we know that there isn't something more effective than mathematics? A prehistoric Wigner might have mused about the unreasonable effectiveness of magical incantations in understanding reality. Was there anything more effective than mathematics that we have not conceived of as yet by definition? Was Wigner asking about the unreasonable effectiveness of whatever we had found to be the most effective?

Indeed, I suspect that what Wigner means by mathematics may just be the entire world of quantitative notions. If so, then the question he raised may be profoundly trivial (or trivially profound). Of course, the bag of quantitative notions is more effective than the bag of qualitative notions, if only because quantitative notions are more precise and compact.

How do we measure reasonableness? According to some sort of community standards? Evidently, since physicists assume that space and time are continuous, and since the assumption is based on well-verified observations, it stands to reason that the mathematics of continuously differentiable functions would be effective. In this view, since the basic mathematical concepts, such as those of geometry, are abstracted out of our experience of the physical world, mathematics ought to be effective. Proponents of this view point to the apparent fact that those areas of modern mathematics least rooted in "everyday" experience tend to be irrelevant to physics.

Perhaps the most dramatic counterexample to this view is the emergence of the complex number in quantum physics. Why, indeed, should the microscopic world be described by complex numbers? It is really rather mysterious.

Coming back to the effectiveness of differentiable functions in describing the physics of continuous spacetime, I should think that if at some distance scale, we were to discover that spacetime is actually discrete, then mathematics (as understood by Wigner, say) would not be effective at all. Indeed, mathematics is not particularly effective in areas such as lattice gauge theory. Not much can be done besides letting the computers go for it.

I want to move on and spend the most time defining "mathematics". Wigner had devoted two sections to "What is mathematics?" and "What is physics?". Rather than repeat what he said, I want to distinguish between mathematics and, for lack of a better term, arithmetic. This distinction is frankly and intentionally touched with a measure of snobbery. ("What you call mathematics is merely arithmetic to me.") Twenty years after Wigner's article appeared, R. W. Hamming wrote an article[2] titled "The Unreasonable Effectiveness of Mathematics". Period. I didn't care much for Hamming's article, because it seemed to me that Hamming, drawing upon his engineering experience, was mostly talking about what I call arithmetic.

Alright, what is the distinction? I would say that mathematics is whatever a reasonably brilliant physicist, defined for the purpose here as someone significantly smarter than I am, could not work out in a finite amount of time, by following more or less straightforward logic. (I will leave it to you to haggle over how long a finite amount of time is.) Everything else is arithmetic. For instance, I probably could have figured out the properties of the solutions of Legendre's equation. All that stuff about Legendre polynomials is definitely arithmetic. On the other hand, the fact that there are only three cases in which higher dimensional spheres can be mapped non-trivially onto lower dimensional spheres, namely $S^3 \to S^2, S^7 \to S^4$, and $S^{15} \to S^8$, that I call mathematics.

Whether I call something arithmetic or mathematics depends to some extent on how we look at it. If we recognize Legendre polynomials as having to do with representations of the rotation group, that indicates some understanding of the structural properties of rotations. In short, I associate mathematics with structural or global understanding, and arithmetic with computation.

Echoing a fairly widespread arrogance of the physicist, Feynman once said that had God not created mathematicians, physics would have been delayed by about a week. (Had complex numbers not been invented by the 1920s, would the development of quantum mechanics have been delayed by significantly longer than a week?) According to Feynman, physicists would have invented what they needed, and the rest, as far as he was concerned, should not have been bothered with in the first place. Feynman's attitude, of course, represents a long tradition in physics. Until the mid-1970s, I would have been inclined to agree with Feynman, but with the advent of superstring theory, and for about a decade before that, truly profound mathematics had started coming into physics, with an intensity that was last seen with the arrival of group theory into quantum physics. But before these developments, it seemed as if physicists could really follow Feynman and have nothing to do with the "pure" mathematicians.

In a way, Feynman was right. We reached the grand unified theory in the 1970s using a minimal amount of mathematics. Some fairly elementary group theory, that's about it. For Pete's sake, the grand unified theory! The theory that unifies three of the fundamental interactions! We have unravelled a big piece of Nature's innermost secrets about how the world is put together without using anything a mathematician would call mathematics. Indeed, the creators of the grand unified theory, and most of the particle physicists of the 1970s, were very Feynmanesque in their disdain for mathematics. Incidentally, Feynman once told me, while we were watching some show, that fancy-shmancy mathematical physics as applied to physics is not worth "a bottle of piss-water".

I like to think of the development of particle physics since the mid-1970s (say) as breaking the shackles of Feynman diagrams. I believe that Feynman diagrams, with all the brilliant simplicity that they incorporate, had too long an influence, and ultimately an unhealthful influence, on particle physics. In a quantum field theory course I took in graduate school, the professor told us that field theory is defined as the totality of Feynman diagrams. The set of diagrams defines a unitary, analytic, and Lorentz invariant theory. All those manipulations in the standard canonical development, such as commuting field operators, are so fraught with delicacies that they are to be regarded as props needed to derive the Feynman rules. Once the rules are determined, quantum fields are to be thrown away.

In this climate, there was indeed no need to learn any mathematics. This view was swept away by the advent of concepts such as solitons and instantons. Particle physicists had to learn about such fancy-shmancy mathematics such as topology.

The ten-year gap between the understanding of spontaneous symmetry breaking and of solitons can, in my opinion, be attributed to the constraining influence of Feynman diagrams. Even in the 1970s there were many people who prefer to describe spontaneous symmetry breaking as the disappearance of diagrammatic lines into the vacuum. The notion of fields as real, as something that we can knead into twisted lumps, was quite revolutionary.

Ironically, the formalism that came into fore, namely the path integral formalism, was also developed by Feynman. That represents a real tribute to Feynman.

With the shackles of Feynman diagrams broken, Feynman's view on mathematics also started to fade. A younger generation of particle physicists felt increasingly at ease with modern mathematics. There was a fundamental shift in outlook towards mathematics, and with the advent of superstring theory around 1983 or so, the trend has accelerated. Today, much of the research in superstring theory is really research into mathematical structures, of a degree undreamed of by Wigner.

With this brief review of how the attitudes of theoretical physicists have changed over the last thirty years or so, let me now come to some observations about the role of mathematics in fundamental physics. You may have already noticed my restriction to "fundamental physics" in my title. I invented this term some years ago to replace the outmoded "particle physics" (or even worse "high energy physics"). Perhaps somewhat tautologically, I define a fundamental physicist as someone who is interested in discovering something fundamental about the physical world. Fundamental physics and particle physics overlap to a large extent, but neither contains the other. The definition is broad enough to include some condensed matter physicists who are interested in understanding the "global" properties of strongly quantum many-body systems. Were I to maintain that arrogant tone I used in distinguishing mathematics from arithmetic, I could have defined a fundamental physicist as someone who tends to use mathematics rather than arithmetic. Anyhow, let's go on.

I believe that the following is a true and somewhat mysterious fact: deeper physics is described by deeper mathematics. Consider the Schrödinger operator versus the Dirac operator. The mathematical structure underlying the Dirac operator is much richer and deeper than the structure underlying the Schrödinger operator. We would expect so since the Schrödinger equation is an approximation of the Dirac equation. Associated with the Dirac operator is real mathematics.

This point was underlined for me quite strikingly some time ago when a colleague and I were studying a condensed matter physics problem of a non-relativistic electron hopping on a two-dimensional lattice in the presence of quantized magnetic flux. The problem has nothing to do with relativistic physics and the Dirac equation. To determine the energy E of the electron as a function of its momentum p, we have a completely standard and straightforward problem of finding the eigenvalues of some n by n matrix. For all but the simplest cases, one would have to go to the computer and crunch some numbers. Nothing inspiring or mathematical about it. However, suppose we are not interested in the precise function $E(p)$ but in determining the number of zeroes, that is, the number of places in p-space where E vanishes. Around such a point p^*, we can expand and in general E will depend linearly on $(p - p^*)$: schematically $E \simeq a(p - p^*)$. With a suitable shift and scale change in the definition of p, we see that the behavior of the electron in that region of momentum space is described effectively by a Dirac equation. What is remarkable is that the entire mathematical edifice of index theorems and winding numbers can now be brought to bear on finding out how many such p^*s there are.

The point is that to determine the function $E(p)$ or even to determine the locations of the p^*s is a job in arithmetic. These quantities depend on the details of the Hamiltonian. Change the Hamiltonian slightly and we expect the value of a given p^* to shift. In contrast, the mathematics tells us that the number of p^*s is invariant as long as the overall structure of the Hamiltonian remains unchanged. In other words, we have the concept of a topological invariant here. Mathematics is effective in giving us global and structural understanding but not in solving computational problems.

I once remarked that more mathematics is associated with the Dirac operator than with the Schrödinger operator to a conference of philosophers interested in physics. Somebody in the audience objected vociferously, "Just count the number of mathematical papers written on the Schrödinger operator!" he said. The confusion here is between usefulness and beauty, so to speak. The Schrödinger equation is useful to a much larger group of physicists than the Dirac equation. Of course people would have devoted a great deal of energy to unravelling the properties of the Schrödinger equation.

Even if we were to focus on the Schrödinger equation, not all Schrödinger problems are created equal. Consider the Schrödinger problem associated with the Stark effect and the Schrödinger problem of a particle moving on a sphere around a magnetic monopole. The latter is associated with deep mathematics, the former is not. But, some of you may think, the former is at least useful, the latter is not, since the monopole may not even exist. In fact, precisely because of the deep mathematics associated with the monopole problem, it has been of central importance in recent developments in theoretical physics and has popped up all over the map. I will mention only a few examples: Berry's phase, Polyakov's instanton in $(2+1)$-dimensional compact gauge theory (a theory which may be relevant to high temperature superconductivity), Haldane's treatment of the antiferromagnetic spin chain, and the movement of holes in a ferromagnetic background. None of these problems has anything to do with the magnetic monopole *per se* but they all share the same underlying mathematical structure.

My story about finding the zeroes also illustrates a point known to every practicing physicist, namely the importance of asking the right question. There are several connotations to the word "right". Obviously, we want to ask physically relevant questions. But we also want to ask questions for which mathematics is effective.

Perhaps sadly, the importance of asking the right question has diminished in some areas of physics due to the availability of computers. Thus, in the problem mentioned above, one can simply compute everything numerically. To some extent, arithmetic can replace mathematics. But inevitably, arithmetic cannot provide the understanding brought by mathematics.

In the thirty years since Wigner's article, we have seen the computer become a major force in theoretical physics. The computer has extended, in a sense, the very domain of theoretical physics. Such fields as chaos and nonperturbative quantum chromodynamics would have been essentially impossible without the computer. At the same time, the computer has pronounced on the subject of Wigner's article: mathematics is not particularly effective in physics, if we define physics as the collection of problems and situations considered by the community of physicists.

An important role played by mathematics is in limiting the possibilities physicists have to consider. An example is the exhaustive classification of Lie algebras. This is obviously of great importance in the development of grand unified theories, for instance. I think that Feynman is wrong here about how physicists can just invent the mathematics that they need. I feel that physicists can probably work out the theory of a specific group, SU(5), say, but the reasoning that allows one to say "Here are all the possible Lie algebras and groups, folks!" is peculiarly mathematical. Actually, the reasoning involved, once it has been invented of course, is not particularly difficult to follow, but it carries that peculiar quality known as mathematical insight.

Of course, there are physicists around, some of the young string theorists, for example, who probably could have worked out the complete classification of Lie algebras. But then these people could have easily become mathematicians as well. Feynman's crack only makes sense if there are distinct personality types, so that out of two persons, equally intelligent according to some measure, one can only be a physicist, the other a mathematician. From my own observations, I believe that that's true to some extent. Many great physicists would have been hopeless as mathematicians.

In truth, there have been many examples of higher mathematics discovered independently by physicists. There is a well-known story by Molière about a gifted but unschooled writer. When a friend complimented him on his prose, our writer was puzzled until he learned that what he was writing was called prose. A physicist colleague of mine was fond of asking mathematically more sophisticated friends in a mocking tone, "Tell me, have I been writing prose?"

It often happens that a discovery in physics is actually associated with a wealth of mathematical structure undreamed of at the time of the discovery. (This is the point I made earlier about the magnetic monopole problem.) Dirac certainly could not have imagined all the mathematics associated with the Dirac operator. A wonderful example in particle physics is the chiral anomaly. It was first discovered in the late 1960s as an oddity: an explicit Feynman diagram calculation had shown that an alleged theorem derived by naively manipulating quantum fields was incorrect. (In fact, already starting in the early 1950s various people had stumbled upon the chiral anormaly in one form or another without recognizing it.) Over the last twenty some years, the chiral anomaly has had a totally amazing habit of popping up in connection with all sorts of major theoretical developments. These developments include the renormalizability of gauge theories, path integral measures, instantons, fractionization of quantum numbers, induced proton decay, winding numbers and intersection numbers, selection of a suitable superstring theory, just to mention some examples. The reason for this remarkable ubiquity is that the chiral anomaly turned out to be associated with a deep mathematical structure rooted in topology and geometry.

It does not follow, however, that objects of great interest to physics is necessarily associated with deep mathematical structures. Consider quantum field theories. In the modern view, a quantum field theory is defined by some sort of functional integral of an integrand equal to the exponential of the action times the imaginary unit i. Apparently, it is just some functional integral out of many possible functional integrals. At least thus far, physicists have not discovered any particularly deep mathematics associated with this integral. In contrast, a part of the integrand, namely the action, often has interesting mathematical properties. (An example is the pure Yang-Mills action.) But in all important cases, knowing some properties of the integrand is not of much help in understanding the properties of the integral. This is well known in statistical mechanics, for example.

Mathematics sometimes popped up uninvited in a physical situation. An example is that of an electron moving in a plane under a uniform magnetic field. The lowest energy quantum wavefunctions turn out to have the form $f(z)e^{-|z|^2}$ where $f(z)$ is *any* holomorphic function and where $z \equiv x+iy$ is the complex coordinate on the plane. Laughlin was able to develop an essentially complete theory of the fractional Hall effect by constructing a variational many-body wavefunction out of these wavefunctions. The structural properties of the theory are made apparent by invoking analyticity theorems at every turn. One has to repeatedly argue along the line "Such and such must have this particular form because of analyticity". *A priori*, the fractional Hall effect poses a formidably difficult problem in many-body dynamics and a theory as complete as Laughlin's would appear to be out of the question.

Remarkably, the wavefunction has the above-mentioned holomorphic form only in a certain gauge (out of an infinity of possible gauge choices). Indeed, one might not have the insight to write the wavefunction in terms of z at all (but instead, in terms of the usual x and y). Conceivably, one can still develop the same theory. After all, at every step, the equations can be gauged, transformed and re-written in terms of x and y. The structural properties of the theory would then be totally obscured.

The preceding illustrates the well-known fact that often, using the right representation can be most of the battle.

This brings us to another fact known to all practicing physicists: the effectiveness (should we say reasonable or unreasonable?) of notation in doing physics. At the simplest but yet a profound level, algebra was invented when someone introduced the "notation" of using letters to represent quantities. In doing physics, we all have our favorite notations to the point that we can barely tolerate an unfamiliar notation. The human mind is a creature of habit. We are used to m for mass and T for temperature, and that's that. Some years ago, a distinguished particle physicist used the letter π as an index (for example, for the πth component of momentum). His papers, which are already quite difficult to read, appeared all that more difficult.

I was told that Maxwell used to write out the components of the electric and magnetic fields starting with E for the first component of the electric field, thus E, F, G, H, I and J. (This is why the magnetic field is called H!) Whether or not this story was invented I do not know, but just imagine doing a standard problem in *E&M* using this notation! The introduction of indices is a truly neat trick. It has also been said that the repeated index summation represents one of Einstein's greatest contributions to physics.

In these examples, a better notation represents heightened efficiency in the sense of the accountant. But often a better notation implies a deeper understanding of the subject. For instance, Dirac's bra and ket notation underlines the fact that we do not have to specify the representation of a state vector in Hilbert space.

Indeed, there are entire topics of mathematics that amount essentially to a better notation. Take differential forms, for example. When I was a freshman, John Wheeler decided, as an experiment, to teach the introductory physics course from "the top down". (Thus, relativity and quantum mechanics were discussed first so that classical physics can be obtained as a "trivial" approximation. Incidentally, the experiment was not repeated the next year.) We were taught electromagnetism using differential forms: "indices without indices" (part of this discussion, using "egg-crates", later appeared in a well-known text on gravity co-authored by Wheeler). Needless to say, we were totally mystified. What was worse was that I, and probably others as well, developed a total distaste for differential forms. It appeared to me useless, since in any specific problems, we eventually had to write out the components of the differential form anyway. For years, I resisted differential forms even as several well-meaning colleagues tried to "teach" me the notation. But about seven years ago, when I was working on anomalies in higher dimensions, I suddenly realized that I could not live without differential forms. If you doubt this, just try solving for w in an equation like trace $F^n = dw$ by writing out everything with indices. You will literally drown in a sea of indices. (Here F is the Yang-Mills gauge field 2-form.)

The real advantage of differential forms is not so much that it saves us from writing out an endless streams of indices, but that it makes clear the geometrical character of various physical quantities. For example, in the magnetic monopole problem, the gauge field 2-form allows us to think of the gauge field F as a single geometrical entity. The writing of F in terms of its components, in contrast, requires commitment to a definite coordinate choice and splits a simple geometrical concept (namely, the concept of area) into an unrecognizable mess. As another example, in that problem mentioned in the preceding paragraph, physicists had long worked out, by using arithmetic, what w is for the case when $n = 2$. But the recognition that as a form, (trace F^n) is closed but not exact conveys a truly deeper understanding.

The trouble was that when Wheeler was trying to convince me of the beauty of differential forms I was trying to master such problems as calculating the electrostatic field of a charged disk, in other words, problems that have about as much underlying geometrical structure as stock pricing analysis. There is no way that differential form can manifest its power in problems that call for only arithmetic.

(This suggests another "definition" of arithmetic versus mathematics: the electromagnetic potential around an electric charge = arithmetic, while the electromagnetic potential around a magnetic charge = mathematics.)

In recent years, various topological concepts such as linking and intersection numbers have entered into physics. Again, they can be written compactly and naturally in terms of a differential form with its underlying geometrical properties.

This story illustrates that mathematics is often too powerful for the physics. Differential form is too much for doing electromagnetism. But when you are tickling non-Abelian gauge theories in higher dimensional spacetime, then differential forms become indispensable.

My earlier attitude towards differential form is typical of the practicing physicist: I'm not gonna learn this stuff unless I can use it for something. My attitude towards fiber bundles, for example, remains at that stage. I have yet to encounter a physics problem in which fiber bundles would help me sigificantly, but I have no doubt that I will eventually. The mathematical concept expressed in fiber bundles strikes me as universal and natural and at some point it is going to seduce me for sure.

Fiber bundle provides an example in which it is useful just to know the words. They serve as pegs on which we can hang our physical concepts, so to speak. Often, they work as mnemonics. For instance, in the fundamental problem of a charged particle moving on a unit sphere around a magnetic monopole, words like sections, while of no actual help to us in solving the problem, remind us that the wavefunction is to be solved on separate patches and then joined together by gauge transformations. In recent years, physicists have used homotopy groups by and large in the same way, as mnemonics more than anything else.

Let us go back to the comparison between the Schrödinger and the Dirac operator. What we gave up in going from the Dirac operator to the Schrödinger operator is of course symmetry: Lorentz symmetry is broken down to rotational symmetry. Lately, I have been particularly struck by the awkwardness of non-relativistic equations when compared to their relativistic counterparts. I was trained as a relativistic physicist, but in the last year and a half I have been working on condensed matter physics. At first, my collaborator had to point out to me constantly that I had erroneously written down a relativistic equation. With a sigh, I would trudge through the non-relativistic form. It would invariably turn out to be much more tedious to deal with.

Deeper mathematics is associated with more symmetrical structures. In 1960, when Wigner wrote his article, the laws of the microscopic world looked rather asymmetric. We now know that those laws are merely phenomenological approximations to deeper laws, which are in fact symmetric. Symmetry has turned out to be a central organizing principle in Nature's design. Indeed, the story of fundamental physics in the last quarter of a century or so has been the profound discovery that as we study Nature at ever deeper levels, Nature exhibits ever larger symmetries.

I have told this story in considerable detail elsewhere.[3] Here I will merely emphasize that it does not have to be such that Nature's laws become more and more symmetric at deeper and deeper levels. For instance, there was a perfectly viable theory of the weak interaction in which the phenomenological Fermi theory was due to the exchange of a pair of scalar particles. Nature could have been designed so that the weak interaction would not be connected to the electromagnetic interaction at all.

Indeed, I think that we can raise the question of "the unreasonable effectiveness of symmetry considerations in understanding Nature". Why should symmetry dominate Nature at the fundamental level? Does the very fact that Nature becomes ever more symmetrical imply that there is a design? Einstein once said that the most incomprehensible thing about the world is that it is comprehensible. *A priori*, we could have lived in a chaotic universe whose working is beyond our comprehension. I have speculated on the philosophical issues raised by these questions in a recent article,[4] and so I will concentrate on the relationship between symmetry and mathematics here.

Symmetry and mathematics are closely intertwinned. Structures heavy with symmetries would also naturally be rich in mathematics. And so if it is indeed true that Nature's design becomes more symmetrical as we probe deeper and deeper, mathematics should be ever more effective.

Let me come back to the distinction between arithmetic and mathematics. In a broad sense, this split is mirrored by the split between dynamics and kinematics in physics. The application of fancy mathematics to physics often amounts to the erection of a kinematical framework within which we can ask dynamical questions. Mathematics is then often not particularly effective at this stage, and arithmetic has to be called upon.

As a specific example, I can refer to the recent discussion of, for lack of a better term, what may be called Chern-Simons theory, with its multifarious possible implications for subjects ranging from topological field theory with its connection to string theory and quantum gravity to high temperature superconductivity. The discussion can be wonderfully mathematical with fancy terms like Hopf terms and braid groups bandied about, but when it comes to actually understanding high temperature superconductivity we have to confront a "real-life" physics problem of working out the statistical mechanics of a liquid of particles with fractional statistics. What is the free energy of this liquid? What are its elementary excitations? Does it behave as a superfluid? Fancy math ain't gonna tell us nothin. Only physical insight and arithmetic will.

In the physics community, people are often involved in value judgment, talking about whether the problem so and so has solved is easy or hard. But in deciding whether or not to be impressed by a colleague's work, people tend to be impressed by fancy mathematics. Paradoxically, problems for which fancy mathematics are effective are often kinematical and hence easy. To quote an example perhaps of little physical importance, I recall that in constructing a Chern-Simons theory of membranes, my collaborator and I were guided at every step by the underlying mathematics of Hopf map, and we knew that things must work out in a preordained way (for instance, that quarternions must enter). Anyone who has worked on a physics problem with a heavy mathematical rather than arithmetical component must have had the feeling that the mathematics has a life of its own and can literally pull one along.

I speak of this split between arithmetic and mathematics from experience as I have worked on both types of physics problem. Perhaps somewhat strangely, I am attracted to both arithmetic and mathematics.

Next, I would like to mention a pervasive feeling among theoretical physicists expressed by the noble sentiment that "If the physics I am working on reveals an unexpectedly rich mathematical structure, then the physics must be correct." We all know that there had been some spectacular confirmations of this hypothesis: Einstein's theory of gravity and

Dirac's theory of the electron, for instance.

This argument is now invoked by some string theorists, and it is certainly true that the mathematical structure hidden in some apparently unprepossessing action describing a string is nothing short of incredible. String theory may well be right, but should we buy this argument? There is a nagging feeling among some people it is no coincidence that the structures studied by fundamental physicists are also precisely those structures favored by mathematicians. The string worldsheet is two-dimensional on which complex numbers and hence analytic functions naturally live. The action may be viewed as a conformal field theory, and all sorts of nice mathematics follow. From the point of view of a naive physicist, it would appear natural to study blobs, rather than strings, if we are going to study extended structures at all. Alas, the blob worldthing is a nasty place where no self-respecting analytic function or conformal field would dare set foot. There does not seem to be a decent mathematical structure at all. Does this mean that string theory is right?

Of course, it may be a waste of time to muse about such things. If I have the strength, I ought to be working on string theory instead. But the preceding discussion has brought us to what I call the dartboard theory of theoretical physics. In the mid-seventies, when there was a proliferation of models of the electroweak interaction, a distinguished experimentalist remarked to a group of us theoretical physicists that theorists are just throwing darts randomly, one of the darts is bound to land, and the wrong theories are just forgotten.

All the theorists who heard this remark were of course outraged, and I think rightly so. The textbook description of the development of physics as a competition between theories does not apply, for the most part, to fundamental physics. At any time, there is usually not a choice between theory A and theory B. Rather, the choice is between a prevailing theory and nothing. We do not have the luxury of choosing between string theory and some other theory. Neither was gauge theory competing with some other theory during the 1970s.

Are mathematicians throwing darts randomly at physics? Out of the wealth of structures studied by mathematicians, isn't it reasonable that some of them are bound to be effective in understanding the physical world?

The influx of mathematics into particle physics over the last few years can only be described as a tidal wave. If you have not followed the development of string theory, let me give you a calibration. In 1984, a theoretical physicist who had a comfortable familiarity with such concepts as coset spaces, homotopy groups, homology sequences, and exceptional algebras would have been regarded by his colleagues as mathematically sophisticated. Some four years later, that same person would be despised by string theorists as a hopelessly unschooled mathematical ignoramus.

Is so much mathematics good for physics? I have no idea. The proof is of course in the eating: we will have to see if string theory can explain the world. Meanwhile, the tidal influx of mathematics is perfectly reasonable. In exploring the physics of the Planck scale, physicists are so far removed from any experimental moorings that mathematics can be our only guide, in a way that Wigner could not possibly have imagined.

In connection with the role of mathematics in physics, I am fond of telling the story[3] of Faraday and Maxwell. Because of his up-from-rags background, Faraday had a self-admitted blind spot – mathematics – and he was unable to transcribe his intuitive notions into precise mathematical descriptions. Just the opposite, Maxwell, scion of a distinguished family, received the best education, in mathematics and in everything else, that his era could provide. But before he began his investigations, Maxwell resolved "to read no mathematics on the subject (of electricity) till I had read through Faraday's *Experimental Researches on Electricity*". Indeed, he considered Faraday's mathematical deficiency an advantage. He wrote: "Thus Faraday ... was debarred from following the course of thought which had led to the achievements of the French philosophers, and was obliged to explain the phenomena to himself by means of a symbolism which he could understand, instead of adopting what had hiherto been the only tongue of the learned."

By "symbolism", Maxwell was referring to Faraday's "lines of force". Earlier, Maxwell had said that "the treatises of (the French philosophers) Poisson and Ampère (on electricity) are of so technical a form, that to derive any assistance from them the student must have been thoroughly trained in mathematics, and it is very doubtful if such a training can be begun with advantage in mature years." Well, I am sure that physicists "of mature years" can all empathize with what Maxwell said.

The story of Faraday and Maxwell is interesting particularly because it is not clear what moral it offers. I think that we are agreed that intuition in the grand tradition of Faraday has been of utmost importance in the development of physics. On the other hand, when you are wandering around in Planckland, what intuition can you possibly have? Let us not forget that Maxwell could probably not have been able to derive the propagation of light without using the methods of the French philosophers, namely differential equations. (Mathematics to him, but arithmetic to us.)

I like to close this musing about the effectiveness of mathematics, reasonable or otherwise, by telling another anecdote.[5,6] A lady who knew Einstein in her youth told me that once, on a brilliant spring day she and Einstein walked into a garden blooming with flowers. They stood looking at the scene in silence. Finally, Einstein said, "We don't deserve all this beauty."

Physics is a beautiful subject made all the more beautiful by the effectiveness of mathematics. Is it reasonable to think that we deserve all this beauty?

References

1. E. P. Wigner, *The Unreasonable Effectiveness of Mathematics in the Natural Sciences*, *Commun. Pure Appl. Math.* **13** (1960) 1–14.
2. R. W. Hamming, *The Unreasonable Effectiveness of Mathematics*, *Am. Math. Mon.* **87** (1980) 81–90.
3. A. Zee, *Fearful Symmetry* (Macmillan Publishing Company, 1986).
4. A. Zee, *Symmetry in the Ultimate Design*, to appear in a book edited by R. Kitchener.
5. S. Asker, private communication.
6. A non-physicist friend to whom I told this story immediately interpreted Einstein's reaction in terms of the collective guilt of physicists in connection with nuclear weapons. I prefer to interpret it at a deeper level, in connection with the discoveries of ever larger symmetries in Nature's design.

Bibliography

Books and Monographs

1. H. M. Blalock, Jr., *Theory Construction: From Verbal to Mathematical Formulations* (Prentice-Hall, 1969).
2. S. Boucher, *The Role of Mathematics in the Rise of Science* (Princeton University Press, 1966).
3. G. R. Boynton, *Mathematical Thinking about Politics* (Longman, 1979).
4. P. W. Bridgman, *The Nature of Physical Theory* (Dover, 1936), Chap. V.
5. Committee on Support of Research in the Mathematical Sciences. NAS-NRC, *The Mathematical Sciences* (MIT Press, 1969).
6. P. J. Davis and R. Hersh, *The Mathematical Experience* (Birkhäuser, 1981).
7. P. J. Davis and R. Hersh, *Descartes' Dream: The World According to Mathematics* (Harcourt Brace Javanovich. 1986).
8. T. J. Fararo, *Mathematical Sociology* (Wiley, 1973).
9. G. Feinberg, *Solid Clues: Quantum Physics, Molecular Biology and the Future of Science* (Simon and Schuster, 1985), pp. 107–119 and Chap. 6.
10. R. Feyman, *The Character of Physical Law* (MIT Press, 1965), Chap. 2.
11. M. O. Finkelstein, *Quantitative Methods in Law* (Free Press — Macmillan, 1978).
12. H. Fruedenthal, *The Concept and Role of the Model in Mathematics and Social Sciences* (Reider, 1961).
13. L. O. Kattsoff, *A Philosophy of Mathematics* (Iowa State College Press, 1948), Chap. 16.
14. J. G. Kemeny and J. L. Snell, *Mathematical Models in the Social Sciences* (Ginn, 1962).
15. P. Kitchen, *The Nature of Mathematical Knowledge* (Oxford University Press, 1983).
16. M. Kline, *Mathematics and the Physical World* (Crowell, 1959), Chap. 27.
17. *Mathematics in the Modern World. Readings from Sci. Am.*, ed. M. Kline (W. H. Freeman, 1968).
18. M. Kline, *Mathematics and the Search for Knowledge* (Oxford University Press, 1985).

19. I. Lakatos, *Mathematics, Science and Epistemology* (Cambridge University Press, 1978).

20. A. Lotka, *Elements of Mathematical Biology* (Dover, 1956).

21. *Mathematics and Life Sciences*, ed. D. E. Mathews (Springer-Verlag, 1977).

22. H. Morowitz and T. Waterman, *Theoretical and Mathematical Biology* (Blaisdell, 1965).

23. H. Pagels, *The Dreams of Reason: The Computer and the Rise of the Sciences of Complexity* (Simon and Schuster, 1988).

24. A. Rapoport, *Mathematical Models in the Social and Behavioral Sciences* (Wiley-Interscience, 1983), Introduction and Chap. 25.

25. A. Renyi, *Dialogues on Mathematics* (Holden-Day, 1967), Chap. 3.

26. R. Rosen, *Dynamical System Theory in Biology* (Wiley-Inerscience, 1970).

27. M. M. Schiffer and L. Bowden, *The Role of Mathematics in Science* (The Mathematical Association of America, 1984).

28. R. Schlegel, *Completeness in Science* (Appleton-Centry-Crofts, 1967).

29. *Mathematical Models in Biological Discovery*, eds. D. L. Solomon and C. Walter (Springer-Verlag, 1977).

30. J. L. Synge, *Science: Sense and Nonsense* (J. Cape, 1951), Chap. 2.

31. C. E. van Horn, *A Preface to Mathematics* (Chapman and Grimes, 1938), Chap. XII.

32. H. Weyl, *Philosophy of Mathematics and Natural Science* (Princeton University Press, 1949).

33. A. W. Whitehead, *Science and the Modern World* (Macmillan, 1925).

34. E. P. Wigner, *Symmetries and Reflections* (MIT Press, 1967), Essays 1–5.

35. R. L. Wilder, *Introduction to the Foundations of Mathematics* (Wiley, 1952), Chap. XII.

36. R. L. Wilder, *Mathematics as a Cultural System* (Pregamon, 1981), Chap. VIII, Sec. 3 and 4.

Papers

1. G. Birkhoff, *SIAM Rev.* **11** (1969) 429–469. Mathematics and Psychology.

2. A. Borel, *Math. Intelligencer* **5** (1983) 9–17. Mathematics: Art and Science.

3. F. E. Browder, *Am. Sci.* **64** (1976) 542–549. Does Pure Mathematics have a Relation to the Sciences?

4. F. Dyson, *Sci. Am.* **211** (1964) 129–147. Mathematics in the Physical Sciences.

5. D. A. T. Gasking, *Australasian Psychology and Philosophy* **XVII** (1940) 97–116. Mathematics and the World.

6. N. D. Goodman, *Am. Math. Mon.* **86** (1979) 540–551. Mathematics as an Objective Science.

7. N. Goodman, *History of Philosophy and Logic* **2** (1981) 55–65. The Experiential Foundations of Mathematical Knowledge.

8. J. V. Grabiner, *Am. Math. Mon.* **81** (1974) 354–365. Is Mathematical Truth Time-dependent?

9. R. W. Hamming, *Am. Math. Mon.* **87** (1980) 81–90. The Unreasonable Effectiveness of Mathematics.

10. S. W. Hawking, *Phys. Bull.* **32** (1981) 15–17. Is the End in Sight for Theoretical Physics?

11. L. Iliev, *Russ. Math Surveys* **27** (1972) 181–189. Mathematics as the Science of Models.

12. K. Knopp, *Math. Intelligencer* **7** (1985) 7–14. Mathematics as a Cultural Activity.

13. N. Koblitz and A. Koblitz, *Math. Intellegencer* **8** (1986) 8–16, 25. Mathematics and the External World: An Interview with Professor A. T. Fomenko.

14. S. MacLane, *Mathematics Form and Function* (Springer-Verlag, 1986), Chap. XII, Sec. 9–11.

15. A. Rapoport, *Am. Math. Mon.* **83** (1976) 85–106, 153–172. Directions in Mathematical Psychology.

16. M. R. Schroeder, *Math. Intelligencer* **7** (1985) 18–26. Number Theory and the Real World.

17. J. T. Schwartz in *Proceedings of the 1960 International Congress on Logic, Methodology and Philosophy of Science* (Stanford University Press, 1962). The Pernicious Influence of Mathematics on Science.

18. C. Smorynski, *Math. Intelligencer* **5** (1983) 9–15. Mathematics as a Cultural System.

19. E. Snapper, *Am. Math. Mon.* **86** (1979) 551–557. What is Mathematics?

20. E. Snapper, *Math. Intelligencer* **3** (1981) 85–88. Are Mathematical Theorems Analytic or synthetic?

21. M. Thompson, in *Mathematics Tomorrow*, ed. L. A. Steen (Springer-Verlag, 1981) Mathematization in the Sciences, pp. 243–250.

22. E. P. Wigner, *Bull. Am. Math. Soc.* **74** (1968) 793–815. Symmetry Principles in Old and New Physics.

Subject Index